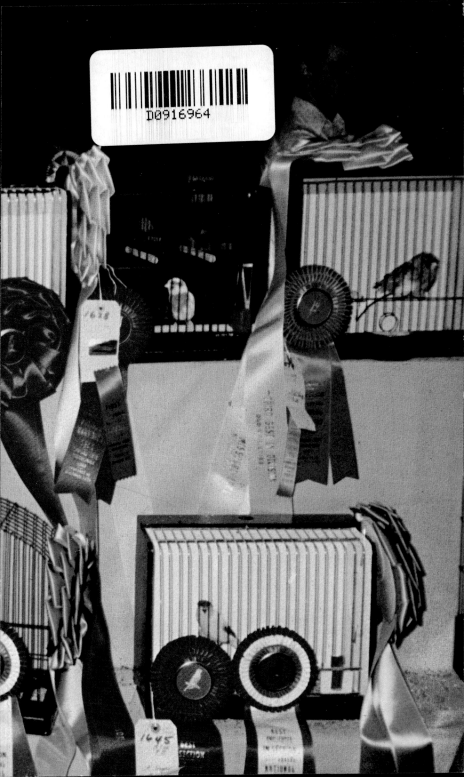

Photography: Courtesy of *American Cage-Bird Magazine*/Arthur Freud, 15 bottom. Glen S. Axelrod, 126 top, 150. Herbert R. Axelrod, 122 bottom, 154. Cliff Bickford, 6, 98, 102, 103, 105, 106, 110, 111, 113, 114, 115, 116, 118, 121, 134, 135, 138, 139, 146, 147, 151, 158, 159, 174, 175, 177, 183, 186, 187. A.E. Decoteau, 87, 119, 142. Pat Demko, endpapers, 10, 11, 14, 17, 18, 19, 22, 23, 26, 27, 66, 162, 166 top, 167, 182 top. Leonard J. Eisenberg, frontis, 30, 31, 32, 40, 67, 70, 71, 74, 75, 78, 79, 82, 83, 84, 100, 166 bottom, 170, 171, 182 bottom. Courtesy of Arthur Freud, 15 top, 36, 37, 38, 44. Michael Gilroy, covers. Earl Grossman, 99, 107. H.V. Lacey, 16, 45, 46, 47, 49, 51, 55, 56, 62, 69 top, 80, 81, 92, 101, 109, 125, 126 bottom, 132, 136, 140, 144, 148, 152, 157, 160, 161, 163, 164, 165, 168, 169, 172, 173, 176, 181. P. Leysen, 123. Gary Lilienthal, 86. A.J. Mobbs, 127, 179. F.B. Mudditt, 63. Stefan Norberg & Anders Hansson, 143. Leslie Overend Ltd., 72. Donald Perez, 90, 91. Laurence E. Perkins, 76. Linda Rubin, 7. Courtesy of San Diego Zoo, 155. Vince Serbin, 53, 64, 145. W.A. Starika, 131. Louise van der Meid, 69 bottom, 89. Courtesy of Vogelpark Walsrode, 50, 122 top, 130, 137.

ISBN 0-87666-830-9

© 1982 by TFH Publications, Inc. Ltd.

Distributed in the UNITED STATES by T.F.H. Publications, Inc., 211 West Sylvania Avenue, Neptune City, NJ 07753; in CANADA by Rolf C. Hagen Ltd., 3225 Sartelon Street, Montreal 382 Quebec; in ENGLAND by T.F.H. (Great Britain) Ltd., 11 Ormside Way, Holmethorpe Industrial Estate, Redhill, Surrey RH1 2PX; in AUSTRALIA AND THE SOUTH PACIFIC by Pet Imports Pty. Ltd., Box 149 Brookvale 2100 N.S.W., Australia; in SOUTH AFRICA by Multipet (Pty.) Ltd., 30 Turners Avenue, Durban 4001. Published by T.F.H. Publications Inc. Ltd., The British Crown Colony of Hong Kong.

EXHIBITING BIRDS

Dr. A. E. Decoteau

Above: Rothschild's Amazon. This individual is in top condition; at exhibitions it has won consistently.

Facing page: As evidenced by this Monk Parakeet at the entrance to her nest, fresh air and the proper environment will produce good condition in a bird that is to be shown.

Dedication

This book is dedicated to Jay Mark Decoteau, my son, who has spent innumerable hours building the aviaries in which our birds have successfully bred, thereby producing a reason for this book.

Contents

Winning Type Canaries. *Above:* Birds
selected by judge Harold Sodamann at the
1981 show sponsored by the National In-
stitute of Red Orange Canaries. *Facing
page:* At the 1981 National Cage Bird Show,
held in Kansas City, the Type Canaries
shown here were placed by judge Margie
McGee.

Acknowledgments

I wish to thank Cliff Bickford for the tremendous photography he has offered once again, for this book—he continues to be a perfectionist in photography. Earl Grossman and Gary Lilienthal are likewise to be thanked for their fine photos. Finally, I am grateful to the Boston Society of Aviculture and Leonard Eisenberg for photographs of the 1980 B.S.A. Show.

Preface

We are in the infancy of an interesting and growing period of exhibitions involving parrots, finches and softbills. Many of the exhibitors today are novices in a growing field. Many questions are asked at the shows; some are answered, some are not. This book, I hope will assist those with an interest in showing and perhaps answer some of their questions. In this book I will stress exhibition of parrots, finches and softbills, with a mention of Canaries and mules or hybrids since these categories are long-time exhibition strongholds. Poultry and pigeons will be ignored as these birds have a separate group of admirers. We are neophytes in a brand-new world of exhibitions, and it is delightful that we can begin on the ground floor. Just think of all the hobbies and avocations that must rely on ancient history. Let us make our own history!

Above: At the 1981 National Cage Bird
Show, the Easterns exhibited by Mary
Hawkins took First in the Rosella Section.

Facing page: The Patagonian Conure shown
by Bruce and Joan Wilson at the 1977
Greater New England Cage Bird Show was
declared Best in Division by judge Arthur
Freud.

Yorkshire canaries staged at a small provincial show in England.

A SHORT HISTORY OF EXHIBITING

WHY EXHIBIT?

For centuries breeders of domesticated birds and small mammals have been interested in comparing the quality of their product with that of others breeding the same breed or variety. Their goal was to selectively breed birds and mammals, often breeding for color, size, feather quality, etc. Competition was keen, so on numerous occasions extensive debates occurred on whether a bird or mammal was superior to the one across the street. Many of these debates occurred at farmers' markets where the product was on display for sale.

Occasionally an owner thought so highly of his product that he refused to sell it, indicating that it was at the farmers' market for exhibition only. Soon his neighbor did the same thing—the debate was on!

In time it was a necessity that an intermediary or judge had to decide which bird or mammal of the same breed was superior. The judge would have to be completely neutral—preferably originating from another city or county, the more distant the better. There could not be any prejudice in such an important decision.

Soon these intermediaries became known for their knowledge of the breed, or variety, of bird or mammal that became their specialty. In due time the demand on them became intense. As interest in selection and production of

Above: Colorbred Canaries placed by judge
Bill Henderson at the 1981 N.I.R.O.C. Show.

Facing page: The 1981 National Cage Bird
Show. *Above:* In the Colorbred Lipochrome
Division, Wallace Graham received a Second, Pat Demko a Third, and Otto Mahnke a
Fourth for their entries. *Below:* Pat Demko,
canary breeder, exhibitor, and judge, poses
with Ignacio Perea, who holds an entry that
took a Third in a Colorbred class.

various specialties grew, so did individuals become specialized judges.

The interest developed from a casual pastime to the point that a group of birds was split into many categories according to likes and dislikes. People started selecting for color, size, feather structure and general beauty, rather than for egg and meat production only. Consequently, numerous beautiful breeds of poultry were developed. Exhibitions at farmers' markets soon developed into fairs and then into giant exhibitions with much competition. Individual breeders wanted to show off the outstanding progeny that they developed through good breeding and selection.

Thus developed perhaps the main reasons for exhibiting birds: to show the offspring of adults you the exhibitor paired and to see if these offspring indeed are better in quality than the adults. Many times a serious exhibitor would be depressed and disappointed with the production of inferior progeny, but often the happy exhibitor would be rewarded by reaping wins with progeny he produced through accurate selection.

By the late 1700's and early 1800's, there were numerous interesting exhibitions in England and Scotland, among other lands, with strong competition. Perhaps the largest and most popular exhibitions involved poultry and dogs. In the United States, the era of exhibitions began to a large extent in the late 1800's.

Today dog exhibitions can be found literally every weekend of the year. Poultry exhibitions appear to be a bit more seasonal, mostly in the spring and fall, with a few occurring at other times. Cagebird exhibitions developed gradually and slowly. English exhibitors made greater strides in cagebird exhibitions than those in the United States, but as cagebirds were increasingly bred, an increase in exhibitions was seen even here.

Canary exhibitions became important throughout the country. With the advent of many Canary clubs for popular varieties such as the Gloster, Border, Yorkshire, Lizard and Norwich as well as the colorbred Canaries, the shows increased in popularity as well as size. Specialty organizations such as The American Budgerigar Society created a keen interest in breeding and exhibiting this popular and fine parakeet. With the advent of the 1970's, an increase in the breeding of rare finches and parrots, as well as those not so rare, created a stir to exhibit in these categories. Cockatiel and lovebird exhibitors developed their standards and soon could present specialty shows to be rivaled by few other types of cagebirds. The rush is on for better selectivity and for better type birds so they will better conform to the standards of perfection. What a thrill it is to win—one can feel those butterflies already!

CAGEBIRD EXHIBITIONS IN THE PAST

Explorers and ornithologists the world over such as Finsch, Jardine, Salvin, Muller, Gould, Barraband and Spix—and even Columbus—not only studied various faunas (including birds) in tropical areas but also brought back many of these specimens. In the early years of exploration these elegant and colorful specimens were presented to royalty. It is known that Queen Victoria was presented with a superb African Grey Parrot. Other members of royalty received Cuban Amazons as early as the 1700's.

As travel to tropical areas became more routine, other people of the upper classes of Europe soon became recipients of these cagebirds. Some, like Dr. Greene, renowned as an author on cagebirds, attempted to breed them, with some success.

In the early 1800's a few cagebird exhibitions occurred, mostly with imported birds, however, and many involved varieties of Canaries produced and improved by selective breeding.

Reports from the middle 1800's describe some large exhibitions with classy psittacines and softbills we seldom see

Above: The winners of the Hookbill and Foreign Division at the 1981 N.I.R.O.C. Show were selected by judge Val Clear.

Facing page: The 1981 National. *Above:* Winners of the American Singer Division, judged by Wayne Vipond, show the rosette color convention: blue for First, red for Second, yellow for Third, and green for Fourth. *Below:* LaVerne Krauss and Mr. Krauss holding the trophy and rosette won by their Second Best Young American Singer.

exhibited today. One show in Manchester, England, was proud to exhibit over twenty Cuban Amazons as well as that beautiful parakeet from America then called the Carolina Conure. We note that the last known living Carolina Parakeet died in 1914 at the Cincinnati Zoo, and today the Cuban is so rare that a repeat of that Manchester exhibition is not likely ever again.

Supply and demand for Canaries and other birds led to major competition and increases in exhibiting. People enjoyed the discussions concerning methods of feeding and took pleasure in preparing standards to be used in producing the ideal exhibition bird. Trading occurred among the exhibitors, always with the thought in mind that new bloodlines for outcrossing might produce better offspring next year. When one show season finished, plans were already underway to prepare to breed birds that would be next year's show winners.

In the late 1890's a few exhibitions were held in the United States, though most at this time were entirely of Canary culture. By the early 1900's, however, the new "parakeet" craze in Europe and the United States began to be cultivated. This parakeet appeared to be an ideal bird to have as a pet as well as to breed. Soon new mutations began occurring, which stirred even more interest. By 1920, this parakeet, which is properly called a Budgerigar, appeared at exhibitions. Breeders worked for new color combinations in the United States, while in England they strived for size. Size they did secure when a new mutation appeared, a much larger bird but still with many color variations. Little did they know at the time that the Americans would covet that larger, more typey exhibition bird over the smaller American budgie.

As the years rolled by, the English shows got larger. American shows also thrived but lacked variety in the early years; Canary exhibitions were dominant for decades. With the importation of English budgies, exhibition of this breed

The top winner at a New England Budgerigar show in 1957 was, of course, from English stock.

increased. One seldom sees American budgies exhibited at a cagebird show today. The exhibition-type English birds would most definitely defeat the American strains by classiness alone, not to mention size, stamina and condition. One also sees more Gloster, Border and American Singer Canaries exhibited, as well as colorbreds, rather than Columbus Fancies, Norwiches, Frills and Yorkshires.

With the exception of Canary and Budgerigar exhibitions, all other kinds of cagebird shows were literally unheard of until the early 1970's. The showing of exotics, namely psittacines (parrots), softbills and finches, is just beginning. But parrots have made tremendous advances as exhibition birds since 1975. At the New Hampshire Cagebird Show, known for its large entry of exotics, the number of parrots went from 26 in 1975 to 52 in 1976, 76 in 1977 and way up to 212 in 1978. That increased interest leads us to believe that the 1980's will be a fantastic decade

Above: The Border exhibited by Paul Dee took Second in Division at the 1981 National, judged by Bernard Lince. *Facing page:* At the N.I.R.O.C. Show a week later, Paul Dee won again with another of his Borders, which received Third Place in the Type Canary Division.

for cagebird exhibition. With the advent of breed specialty associations for Cockatiels and lovebirds, a grand increase in exhibiting has been noted. The National Cage Bird Exhibition, with a total entry of over 1,500 birds, can boast of a tremendous class of 165 Cockatiels alone. The time has come for competitive action. Try it!

SHOW TRENDS

A good example of these can be found in the history of The National Cage Bird Exhibition, held every November in a different area of the United States. In Texas in 1977, there were about 20 Cockatiels exhibited, with perhaps 70 to 80 parrots in total. During this time the American Cockatiel Society was becoming well organized and had decided to conduct a Cockatiel specialty in conjunction with the 1978 National in Georgia. At least 125 Cockatiels were exhibited, with Nancy Reed of Connecticut taking Best Cockatiel in the show. In 1979, the National show was held in Los Angeles, California. Again, the American Cockatiel Society had a specialty with approximately 85 cockatiels exhibited. Nancy Reed was again winner of Best Cockatiel, for the fifth consecutive year.

In 1978, when the National was held in Georgia, the lovebird entry was poor, with not over ten exhibited. That year The African Lovebird Society, through their superb magazine *Agapornis World*, outlined plans to hold a specialty show in Los Angeles in conjunction with the 1979 National. A thrilling group of 130 lovebirds showed, with an Abyssinian winning Best Lovebird.

In 1980 in Florida, the author was fortunate to judge the lovebird entry; there were 50 to 60 classy lovebirds entered. Ohio's Dr. Richard Baer, of American Federation of Aviculture (AFA) fame, won Best Lovebird with a Lutino Peach-faced. At the same show Nancy Reed judged a record entry of Cockatiels, placing the normal male entered by Dee Dee Squyres of Texas over 156 others.

Larger parrots are exhibited, but still not in great numbers. Even in England, where the shows are tremendous in size, it is notable that the number of large parrot entries is low. Perhaps the biggest parrot show of recent years I can recall was the New Hampshire Cagebird Show of 1978. This show is held on the second weekend of October each year. In 1978, its record year, there were 212 parrots and close to 100 finches exhibited.

England, of course, still takes the lead in exhibitions, although I must disagree with their system of classifying both parrots and finches. However, their shows are spectacular, the Canary and Budgerigar entries are fascinating, and they excel in mules and hybrids.

Show entries may vary from season to season. For instance, due to an Exotic Newcastle scare in late September and early October of 1980, many shows indicated tremendous reductions in entries. New Hampshire Cagebird was down to 20 parrots and 64 finches; the Boston show had 30 parrots but only 19 finches; and the very old and extraordinary Massachusetts Cagebird Show in Plymouth had about 15 parrots entered. As the fall progressed, more and more parrots and finches were evident at the shows, however. Things are almost back to normal now, and perhaps within the next four or five years some show, perhaps the National, will have achieved the 500-mark for parrots as well as finches.

The 1980 Boston Society for Aviculture
Show. *Above:* Staging of some of the entries in the show hall in Weston, Massachusetts. *Facing page:* Behind the rosettes for
the Parrot Division, judge Mark Runnals
(*left*) talks with Richard Mather, probably
about parrots.

Publicity poster for the 1980 B.S.A. Show.

SHOWS AND THEIR ORGANIZATION

THE CLUB AFFILIATION

Just as the farmers' markets and fairs in which bird exhibiting has its roots were local or regional in character, most shows today are sponsored by clubs whose members live in the same vicinity. The members of these general-interest clubs fancy different birds, so their shows will have divisions to accommodate all kinds of birds, like the Massachusetts Cagebird Association, for instance.

Whenever some cagebird becomes increasingly popular as a pet, for breeding and for exhibition, then people are aroused into forming an organization for the purpose of further developing the kind of bird in which they have the deepest interest. Some of these specialty groups have been in existence for many years; one such organization is the National Gloster Club.

Most specialized clubs are national, but some areas have enough people interested in birds that regional specialty clubs are formed as well. On the other hand, few general-interest associations are national in scope; two that are, though, are The American Federation of Aviculture and The National Cage Bird Show Club. Two of the newest associations have made great strides by becoming almost international in scope. The American Cockatiel Society and The African Lovebird Society are strong in membership, having developed impressive national shows. Both organizations publish a national bulletin of excellent caliber. *The American Cockatiel Society Bulletin* and *Agapornis World* (the lovebird journal) are filled with treasures of material about these birds.

Noted that as increased interest is aroused and the organizations grow, many new and beautiful mutations appear in the birds. This in turn brings out more entries and greater competition. The 1979 National Show in Los Angeles brought out over 80 Cockatiels and 130 lovebirds to compete for the splendid and coveted Kellogg trophies. Of course, in addition to exhibiting to seek for proof of good, selective breeding, that trophy and rosette and the cash award are also attractions.

Other national associations worth mentioning are: The American Border Fancy Canary Club; The American Budgerigar Society; The American Dove Association; The American Norwich Society; The American Singers Club; The Yorkshire Canary Club of America; The U.S. Association of Roller Canary Culturists; and The Avicutural Society of America. There are many others that meet periodically, with several goals in mind. The greatest goal, however, is that yearly exhibition. Who will win that Best in Show?

FEATHER SHOWS

For many aviculturists, the breeding seasons of their birds occur shortly after the show season ends. People are by then mating their pairs of Canaries, finches, budgies and parrots to ensure good-quality youngsters for the coming season. But the spring and summer months do not have to be devoid of the enjoyment of exhibiting birds. Numerous people belong to cagebird associations that conduct a summer "feather show" in conjunction with an annual picnic or some other function.

All of the new hatchlings are by then growing their new feathers and developing into adults. They are coming into feather! The owners enjoy showing off their young birds in

One of the entries checking his classification in the New Hampshire Cage Bird Show catalog.

these summer shows. Another name for these new youngsters, a name which one sees utilized frequently, is "unflighted." This is really a fun show, with ribbons, rosettes and trophies awarded to the best young birds selected as winners. Judges are generally local talent or apprentice judges seeking the experience. Most of the feather shows include mainly local members of the participating cagebird association, while large fall shows bring entries from many states. Feather shows frequently take place in June, July or August.

The New Hampshire Cagebird Association has conducted a feather show annually in June or July for several years. It is always conducted as an outdoor show. Entries are most interesting. It has also rained every single year, but fortunately the judging has been completed before the rains developed.

Feather shows are good for novice exhibitors and breeders. They assist them in preparing their birds for ex-

1981 Long Island Cage Bird Association Show: Arthur Freud, Parrot and Cockatiel judge, confers with Natalie Molaver, judge's secretary, and Russell Armitage, assistant steward, before conferring the awards.

hibition, ensure them time to see just how their birds are judged, and give them an opportunity to discuss birds with other breeders and exhibitors. Larger, more sophisticated shows are busy in many ways, often to a point where the novice finds people too busy, hurrying from one thing to another, to answer questions. Judges are busy with their assignments, and when they are completed there is a long line of people waiting to talk with them. In a feather show the general routine is simpler and more easy-going. This is truly where the novice should begin. Feather shows are much fun!

ORGANIZING A SHOW

The show season for cagebirds usually starts in early October and continues every weekend until about the middle of December. A few shows take place in January, and there is an occasional specialty show in September. Regional cagebird associations start planning their all-cagebird shows as early as November or December of the previous year. The next year's show date and location must be considered early. Often the show chairman for the next year is elected quite early so he or she can have sufficient planning time.

By late winter most association directors and show chairmen have already secured bids from various judges throughout the country and the world for the purpose of binding the specific dates. Often the outstanding judges are booked up quite fast. The New Hampshire Cagebird Association, for instance, frequently employs as many as nine judges in the following categories: Parrots, Finches and Softbills; Cockatiels (as a specialty show within the general club show); Budgerigars; Colorbred Canaries; Gloster Canaries; Border Canaries; Type Canaries (which includes Yorkshires, Norwiches, Lizards and Frills) and Mules and Hybrids.

1981 Long Island Cage Bird Association Show: Canary judge Tony Munoz taking a breather.

Show stewards are responsible for staging the birds during judging, and many other vital tasks as well. At the 1981 L.I.C.B.A. Show, Marc Marrone stewarded the Parrot and Cockatiel Division.

Once the judges have been selected, the show chairman has many committees to activate. The cost of holding a cagebird show is increasing every year; I recall that the 1979 New Hampshire show cost well over $2,500. Therefore catalog advertising must be solicited. Many potential exhibitors, allied industries and aviculturists donate funds toward trophies and rosettes. The chairman must set up a trophy committee as well as a ribbon and rosette committee.

Trophy selections as well as rosettes must be appealing to exhibitors. I have seen a great difference in number of entries between shows presenting flimsy, poor-looking trophies and shows that present gigantic, quality trophies. In the long run, the shows with better trophies reap more entries of birds for the next year. Of course, not all exhibitors care for trophies. Some return them to the show committee for future presentation, and some indicate that they are dust collectors. There are some show committees that present sterling-silver trophies, while others present victory trophies. More and more, some clubs are utilizing bird figures on their victory trophies.

The show chairman must select someone who is knowledgeable to prepare the schedule of classes for the catalog. This will be discussed in a later section in greater detail.

Each judge should have a secretary and a show steward. The steward can either make or break a show. He facilitates judging, because a good steward will have entries ready at the judging table for the judge to view. By the time one class is completed, a top steward has all entries at the judge's table ready for the next class. A good steward doesn't miss any entries. I recall that at one show at which I officiated the steward missed one important entry that was not judged. This was embarrassing to everyone and produced an irate exhibitor. A steward must be trained in advance and certainly should know the identities of birds in the classes which he is stewarding. This is one reason I like to look at all birds to be judged prior to the beginning of judging.

Likewise, a judge's secretary is important. He or she should tend to the business at hand so as to record *every* decision made by the judge. I also recall one secretary who missed recording several special awards. This created a terrible situation until it was finally corrected.

There are numerous other committees that make or break a show: the entry-taking committee, the gate committee (so the general public can pay to see the birds), the door-prize committee (often shows award special birds to visitors and exhibitors as an added attraction), the set-up committee (important for obtaining benches and tables for the exhibits, plus taking care of a hundred other details) and, finally, a clean-up committee (if the association leaves a dirty hall, you have to search for another site the next year). Obviously much behind-the-scenes work is necessary in developing a good show.

A Border canary winner at the 1980 B.S.A. Show.

CLASSIFICATION FOR SHOWING

Most bird clubs and avicultural associations wind up their year by holding a bird show, generally in the fall of the year. Their toughest project is to prepare the show catalog, which sets out the classes of competition and the awards that will be presented.

Grouping birds into different classes for competition varies considerably, particularly with parrots and finches. Some clubs have poorly planned classifications, while others are more successful; yet none appear to be perfect. But fairness in competition should be the most important consideration. Some show committees will group lovebirds, parrotlets and ringnecked parakeets together, which is about like judging an apple against a pear and a banana. I believe that each entry should first be compared with others of the same species or variety, and then compared to the ideal for that kind of bird. The Best award is given to the bird that conforms most closely to that standard.

One good classification scheme for parrots is based on taxonomic and morphological features; another utilizes geographical distribution. I believe that a combination of the two is necessary, such as the one employed by Forshaw in his *Parrots of the World*. No author in history has so successfully catalogued the parrots.

Certainly there are some rare groups that must be combined in All Other classes, but those parrots most often exhibited should compete only against others of the same species. For competition, then, the order of parrots is best classified into four major groups: South Pacific, Afro-Asian, South American and the Mutations. The South Pacific group includes all parrots from Indonesia through the South Pacific, including New Guinea, Australia, Tasmania and New Zealand. The Afro-Asian group includes parrots native to Africa and Asia, excluding Indonesia. South American parrots encompass all South American species and also those from Central America, North America and the Caribbean. The fourth group includes all mutations, including those which have been successfully developed by man's manipulation through selective breeding. This group has increased in numbers recently, literally tripling during the 1970's. For example, in 1970 little more than one or two varieties of the Peach-faced Lovebird were occasionally evident. Today many more are available, and we can expect others in the future.

As captive breeding of a species continues, more and more varieties are developed. As with lovebirds, new Cockatiel mutations are appearing at shows. As the specialty grows, classes must be set up for Lutinos, Pearls, Cinnamons and Pieds. The Any Other class will include even newer varieties such as Pearl Pieds, Cinnamon Pieds and Cinnamon Pearls. Someday we can expect some of these birds to require classes of their own. In the meanwhile, the entry of Normal Cockatiels will become sufficiently numerous that they will compete in different classes for young cocks, young hens, old cocks and old hens.

The most important determinant in preparing the schedule of classes is an estimate of what birds are likely to be entered. The many varieties of Canaries and Budgerigars and their popularity among fanciers means that the entry in these divisions will be large and will require numerous classes. Most organizations structure their schedules so that *classes* are grouped into *sections*, which are

42

parts of *divisions*. Divisions are set up for all the major kinds of birds (parrots, finches, Canaries and Budgerigars, for instance), depending on their popularity as reflected in show entries.

At many shows today separate divisions are needed for lovebirds and Cockatiels. In other cases there are so few birds exhibited that it would be senseless to set aside an entire division for a very small entry (the cost of rosettes, ribbons and trophies for one division is enormous). A good example of this is doves; they are frequently placed in a division with softbilled birds. The entries at some shows dictate including doves, softbills and finches in the same division.

The arrangement of the schedule of classes must indicate the levels of competition and reflect the awards offered. In my classification of parrots, for example, the winners of each class of cockatoo will compete against each other for Best Cockatoo. Best Cockatoo then competes against Best Lory, Best Rosella, Best Grass Parakeet, etc. for Best South Pacific Parrot. The Best South Pacific Parrot stands against the Best Afro-Asian Parrot, the Best South American Parrot and the Best Mutation Parrot for the Best Parrot in the show.

Finches also compete by species in various geographic groups. Asian, African, Australian, European and South American finches are in different sections. The Best from each section will compete for the division award, Best Finch.

While a class-section-division arrangement is adequate for the finches, Canary classifications are more complex. In the Gloster fancy, for example, coronas and consorts are placed in separate sections, with classes distinguished according to sex and age, as well as other factors. Eventually a Best Gloster is selected. It competes for Best Canary with the winners from the other major groups (subdivisions): Best Border, Best Colorbred, Best Type, etc.

Show officials have the important job of making sure entries are appropriately classified. At the 1981 L.I.C.B.A. Show, an entry held by show manager Gabe Dillon (*left*) and show secretary Richard Norrby (*center*) is discussed.

Budgerigars are similarly popular, and classes are numerous. Different classes are set up for cocks and hens, both young and old, for almost every color. In Exhibition Budgerigars, competition is further distinguished into Novice, Intermediate and Championship. Thus Budgerigars have a division to themselves. The Best of this division competes against the other division champions for Best in Show. When it is time to select the Best in Show, usually several judges debate and select the winner. While this selection may be made in open debate, it is best done by closed ballot, each judge assigning ten points to his winning selection, five points to the second, and one point to the third. Compilation is done by the show chairman, who announces the winner of Best in Show.

The following model schedule illustrates a classification that might be used in a current show catalog. Many variations are possible, of course, to accommodate particular cases, but the arrangement follows the recommendations outlined above.

SCHEDULE OF CLASSES

Awards: *Best in Show*
              1st, 2nd & 3rd in each class

DIVISION 1 — PARROTS

Award: *Best Parrot*

SUBDIVISION 1 — SOUTH PACIFIC PARROTS

Awards: *Best South Pacific Parrot*
              2nd & 3rd Best South Pacific Parrot
              Best, 2nd & 3rd in each section

Section 1—Lories
1101   Chattering
1102   Red
1103   Dusky
1104   Black-capped
1105   Rainbow
1106   Scaly-breasted
1107   All Other Lories

Section 2—Cockatoos
1201   Greater Sulphur-crested
1202   Lesser Sulphur-crested
1203   Salmon-crested
1204   Galah
1205   White
1206   All Other Cockatoos

Section 3—Rosellas
1301   Eastern
1302   Pale-headed
1303   Crimson
1304   Western
1305   All Other Rosellas

Eastern Rosella

Section 4—Grass Parakeets
1401   Bourke's
1402   Turquoise
1403   Scarlet-chested
1404   All Other Grass Parakeets

Section 5—All Other South Pacific Parrots
1501   Red-rumped
1502   All Other South Pacific Parrots

SUBDIVISION 2 — SOUTH AMERICAN PARROTS

Awards: *Best South American Parrot*
                2nd & 3rd Best South American Parrot
                Best, 2nd & 3rd in each section

Section 1—Macaws
2101   Scarlet
2102   Blue-and-Gold
2103   Green-winged
2104   Military
2105   Hyacinth
2106   Chestnut-fronted
2107   Yellow-collared
2108   All Other Macaws

Section 2—Conures
2201   Orange-fronted
2202   Green
2203   Nanday
2204   Mitred
2205   Jandaya
2206   Sun
2207   Peach-fronted
2208   Patagonian
2209   Maroon-bellied
2210   All Other Conures

Maroon-bellied Conure

Section 3—Amazons
2301   Double Yellow-headed
2302   Yellow-fronted
2303   Yellow-naped
2304   Mealy
2305   Blue-fronted
2306   Orange-winged
2307   Red-lored
2308   White-fronted
2309   Plain-colored
2310   Hispaniolan
2311   Rothschild's
2312   Green-cheeked
2313   All Other Amazons

Section 4—Pionus
2401   Blue-headed
2402   Slaty-headed
2403   White-capped
2405   All Other Pionus

Section 5—Caiques
2501   Black-headed
2502   White-bellied

Canary-winged Parakeets

Section 6—South American Parakeets
2601   Tui
2602   Monk
2603   Barred
2604   Canary-winged
2605   All Other South American Parakeets

Section 7—Hawk-headed Parrot
2701   Hawk-headed

Section 8—All Other South American Parrots
2801   All Other South American Parrots

SUBDIVISION 3 — AFRO-ASIAN PARROTS

Awards: *Best Afro-Asian Parrot*
              2nd & 3rd Best Afro-Asian Parrot
              Best, 2nd & 3rd in each section

Section 1—Lovebirds
3101   Peach-faced
3102   Fischer's
3103   Masked
3104   Abyssinian
3105   All Other Lovebirds

Section 2—African Grey Parrots
3201   African Grey
3202   Timneh African Grey

Section 3—Ringnecked Parakeets
3301   Rose-ringed
3302   Moustached
3303   Alexandrine
3304   Plum-headed
3305   All Other Ringnecked Parakeets

Section 4—All Other Afro-Asian Parrots
3401   All Other Afro-Asian Parrots

SUBDIVISION 4 — MUTATION PARROTS

Awards: *Best Mutation Parrot*
              2nd & 3rd Best Mutation Parrot

Section 1—All Mutation Parrots
4101   American Pied Peach-faced Lovebirds
4102   Yellow Peach-faced Lovebirds
4103   Dutch Blue Peach-faced Lovebirds
4104   Blue Masked Lovebirds
4105   Lutino Rose-ringed Parakeets
4106   Blue Rose-ringed Parakeets
4107   Yellow Red-rumped Parakeets
4108   Yellow Bourke's Parakeets
4109   All Other Mutation Parrots

Senegal Parrot. In the Schedule of Classes given in these pages, the Senegal would be entered in Class 3401: All Other Afro-Asian Parrots.

DIVISION 2 — FINCHES

Awards: *Best Finch*
     2nd & 3rd Best Finch
     Best, 2nd & 3rd in each section

Section 1—Australian Finches
101 Gouldian Finch
102 Star Finch
103 Chestnut-breasted Mannikin
104 Long-tailed Grassfinch
105 Masked Grassfinch
106 Parson Finch
107 Owl Finch
108 Sydney Waxbill
109 Normal Zebra Finch
110 White Zebra Finch
111 Fawn Zebra Finch
112 Chestnut-flanked White Zebra Finch
113 All Other Zebra Finches
114 All Other Australian Finches

Owl Finch

Section 2—Asian Finches
201 Strawberry Finch
202 Black-headed Nun
203 White-headed Nun
204 Tricolored Nun
205 Spice Finch
206 Striated Munia
207 Indian Silverbill
208 Bengalese Finch
209 Pintailed Nonpareil
210 Normal Java Sparrow
211 White Java Sparrow
212 Pied Java Sparrow
213 All Other Asian Finches

Section 3—South American Finches

301   Black-crested Finch
302   Cuban Finch
303   Saffron Finch
304   White-throated Finch
305   Crimson-pileated Finch
306   Jacarini Finch
307   All Other South American Finches

Section 4—African Finches

401   Green Singing Finch
402   Gray Singing Finch
403   Gold-breasted Waxbill
404   Red-eared Waxbill
405   St. Helena Waxbill
406   Orange-cheeked Waxbill
407   Lavender Waxbill
408   Common Fire Finch
409   Cordon Bleu
410   Violet-eared Waxbill
411   African Silverbill
412   Bronze-winged Mannikin
413   Cut-throat
414   Orange Bishop
415   Napoleon Weaver
416   Red-billed Weaver
417   Paradise Whydah
418   Pintailed Whydah
419   Combassou
420   All Other African Finches

Section 5—European Finches

501   Goldfinch
502   Chaffinch
503   Bullfinch
504   All Other European Finches

Black-crested Finch

DIVISION 3 — SOFTBILLS, DOVES AND QUAIL

Awards: *Best in Division*
          Best, 2nd & 3rd in each section

Section 1—Softbills
101   Mynah Birds
102   Pekin Robin
103   Toucans
104   Toucanettes
105   Rollers
106   Hummingbirds
107   All Other Softbills

Section 2—Doves and Quail
201   Button Quail
202   Diamond Dove
203   Cape Dove
204   Emerald Dove
205   Green-winged Dove
206   Crested Dove
207   Triangular-spotted Dove
208   Nicobar Pigeon
209   Bleeding Heart Pigeon
210   All Other Doves

DIVISION 4 — CANARIES

Award: *Best Canary*

SUBDIVISION 1 — GLOSTERS
Judged to I.G.B.A. Standards

Awards: *Best Gloster*
          Best Corona
          Best Consort
          2nd & 3rd in each section
          Best Young in each section

Toco Toucan. At most bird shows, the number of toucans entered will be small, which increases the likelihood that a specimen in as fine condition as this one will emerge a winner.

Section 1—Coronas

Cocks:		Hens:		
Old	Young	Old	Young	
1101	1102	1103	1104	Buff, clear/ticked, dark corona
1105	1106	1107	1108	Buff, variegated
1109	1110	1111	1112	Three parts dark
1113	1114	1115	1116	Yellow, clear/ticked/variegated
1117	1118	1119	1120	White or White-ground
1121	1122	1123	1124	Cinnamon, self/foul/variegated
1125	1126	1127	1128	Green, self/foul
1129	1130	1131	1132	Clear, grizzled corona

Section 2—Consorts

Cocks:		Hens:		
Old	Young	Old	Young	
1201	1202	1203	1204	Buff, clear/ticked, dark corona
1205	1206	1207	1208	Buff, variegated
1209	1210	1211	1212	Three parts dark
1213	1214	1215	1216	Yellow, clear/ticked/variegated
1217	1218	1219	1220	White or White-ground
1221	1222	1223	1224	Cinnamon, self/foul/variegated
1225	1226	1227	1228	Green, self/foul

SUBDIVISION 2 — BORDERS
Judged by I.B.F.C.C. Standards

Awards: *Best Border*
            2nd, 3rd & 4th in section
            Best Young (Members only)

Section 1—All Borders

Cocks:		Hens:		
Old	Young	Old	Young	
2101	2102	2103	2104	Yellow, clear/ticked
2105	2106	2107	2108	Buff, clear/ticked
2109	2110	2111	2112	Yellow Green, variegated
2113	2114	2115	2116	Buff Green, variegated
2117	2118	2119	2120	Yellow Cinnamon, variegated
2121	2122	2123	2124	Buff Cinnamon, variegated
2125	2126	2127	2128	Yellow Green, three parts dark variegated
2129	2130	2131	2132	Buff Green, three parts dark variegated
2133	2134	2135	2136	Yellow Cinnamon, three parts dark variegated
2137	2138	2139	2140	Buff Cinnamon, three parts dark variegated
2141	2142	2143	2144	Yellow Green, self/foul
2145	2146	2147	2148	Buff Green, self/foul
2149	2150	2151	2152	Yellow Cinnamon, self/foul
2153	2154	2155	2156	Buff Cinnamon, self/foul
2157	2158	2159	2160	White, clear/ticked
2161	2162	2163	2164	White, variegated
2165	2166	2167	2168	All Other Colors

SUBDIVISION 3 — SINGERS

Award: *Best Singer*

Section 1—American Singers
Judged by American Singer Standard Group System, adopted 1967

Awards: *Best American Singer*
            2nd & 3rd in section
            1st, 2nd & 3rd in each group

American Singers will be judged by the Group System. Each bird will be entered in a different group, regardless of color. An exhibitor will not compete against himself in the same group, unless he enters more birds than groups formed. Write band number in full when listing entries on entry blank (initial, number, and year). Only plain canary seed allowed in American Singer cages.

Cocks: Hens (judged for type only):

Old	Young	Old	Young	
3101	3102	3103	3104	All Colors

Section 2—Variety Singers

Awards: *Best Variety Singer*
                    2nd & 3rd in section

Old	Young	
3201	3202	Buff, clear/ticked
3203	3204	Yellow, clear/ticked
3205	3206	Green, self/foul/variegated
3207	3208	Cinnamon, self/foul/variegated
3209	3210	White
3211	3212	Red Factor, clear/ticked/variegated

Cinnamon Border

In terms of Section 3 on the facing page, the show officials will have to determine whether this Frill is a Northern Dutch; if not, it must be entered in one of the All Other Varieties classes.

SUBDIVISION 4 — VARIOUS TYPE CANARIES

Section 1—Yorkshires

Awards: *Best Yorkshire*
                   2nd & 3rd in section
                   Best Young in section

Cocks:		Hens:		
Old	Young	Old	Young	
4101	4102	4103	4104	Yellow, clear/ticked
4105	4106	4107	4108	Buff, clear/ticked
4109	4110	4111	4112	Yellow Green, variegated
4113	4114	4115	4116	Buff Green, variegated
4117	4118	4119	4120	Yellow Cinnamon, variegated
4121	4122	4123	4124	Buff Cinnamon, variegated
4125	4126	4127	4128	Green, self/foul
4129	4130	4131	4132	Cinnamon, self/foul
4133	4134	4135	4136	White
4137	4138	4139	4140	All Other Colors

Section 2—Norwiches

Awards: *Best Norwich*
    2nd, 3rd & 4th in section
    Best Young in section

Cocks:		Hens:		
Old	Young	Old	Young	
4201	4202	4203	4204	Yellow, clear/ticked/variegated
4205	4206	4207	4208	Buff, clear/ticked/variegated
4209	4210	4211	4212	Green, self/foul
4213	4214	4215	4216	Cinnamon, self/foul
4217	4218	4219	4220	All Other Colors
4221	4222	4223	4224	Crested, All Colors

Section 3—Other Type Canaries

Award: *Best in section*

Old	Young	
4301	4302	Lizard
4303	4304	Northern Dutch Frill
4305	4306	Belgian Fancy
4307	4308	All Other Varieties

SUBDIVISION 5 — COLORBRED CANARIES
Judged by N.I.R.O.C. Revised Standards

Awards: *Best Colorbred*
    2nd & 3rd Best Colorbred
    Best, 2nd, 3rd in each section
    Best Young in each section

Section 1—Lipochrome: Nonfrosted Red-orange
(Cocks or Hens)

Old	Young	
5101	5102	Nonfrosted Red-orange, clear/ticked
5103	5104	Nonfrosted Red-orange, variegated

Section 2—Lipochrome: Frosted Red-orange (Cocks or Hens)

Old	Young	
5201	5202	Frosted Red-orange, clear/ticked
5203	5204	Frosted Red-orange, variegated

57

Section 3—Lipochrome: Mosaic (Dimorphic) Hens

Old	Young	
5301	5302	Mosaic (Dimorphic) Hens

Section 4—Melanin: Red-orange Ground (Cocks or Hens)

Frosted:		Nonfrosted:		
Old	Young	Old	Young	
5401	5402	5403	5404	Bronze
5405	5406	5407	5408	Bronze Pastel
5409	5410	5411	5412	Red-orange Agate
5413	5414	5415	5416	Red-orange Agate Pastel
5417	5418	5419	5420	Red-orange Brown
5421	5422	5423	5424	Red-orange Brown Pastel
5425	5426	5427	5428	Red-orange Isabel (Dilute)
5429	5430	5431	5432	Red-orange Isabel (Dilute) Pastel

Section 5—Melanin: Yellow Ground (Cocks or Hens)

Frosted:		Nonfrosted:		
Old	Young	Old	Young	
5501	5502	5503	5504	Green
5505	5506	5507	5508	Green Pastel
5509	5510	5511	5512	Agate
5513	5514	5515	5516	Agate Pastel
5517	5518	5519	5520	Brown
5521	5522	5523	5524	Brown Pastel
5525	5526	5527	5528	Isabel (Dilute)

SUBDIVISION 6 — MULES AND HYBRIDS

Awards: *Best Mule or Hybrid*
              2nd & 3rd Best Mule or Hybrid

Section 1—Mules and Hybrids

Old	Young	
6101	6102	Goldfinch Mules
6103	6104	Linnet Mules
6105	6106	All Other Mules and Hybrids

DIVISION 5 — AMERICAN BUDGERIGARS

Awards: *Best American Budgerigar*
                  Best, 2nd, & 3rd in each section

There is no official standard for judging the American Budgerigar. However, a good guide is the bird's condition. The term *condition* applies to physical condition, health, cleanliness, full plumage, and training. A bird need not be tame to enter into competition. Birds may be exhibited in any cage suitable for Budgerigars. Food and water vessels will be permitted during judging. No names, distinctive marks or decorations will be allowed on cages until after judging has been completed.

Section 1—Normal Greens

Cocks	Hens	
101	102	Light-green
103	104	Dark-green
105	106	All Other Normal Greens

Section 2—Normal Blues

Cocks	Hens	
201	202	Skyblue
203	204	Cobalt or Mauve
205	206	Violet
207	208	All Other Normal Blues (incl. Yellow-face)

Section 3—Opalines

Cocks	Hens	
301	302	Light-green, Dark-green, or Olive-green
303	304	Skyblue, Cobalt, or Mauve
305	306	Violet
307	308	All Other Opalines (incl. Yellow-face)

Section 4—Opaline Cinnamons

Cocks	Hens	
401	402	Light-green, Dark-green, or Olive-green
403	404	Skyblue, Cobalt, or Mauve
405	406	Violet
407	408	All Other Opaline Cinnamons (incl. Yellow-face)

Section 5—Wing Series

Cocks	Hens	
501	502	Cinnamon Green Series
503	504	Cinnamon Blue Series
505	506	Greywing Green Series
507	508	Greywing Blue Series
509	510	Clearwing Green Series
511	512	Clearwing Blue Series
513	514	All Other Wing Series (incl. Yellow-face)

Section 6—Miscellaneous Rares

Cocks	Hens	
601	602	Lutino
603	604	Albino
605	606	Fallow
607	608	Pied Green
609	610	Pied Blue
611	612	Opaline Pied Green
613	614	Opaline Pied Blue
615	616	Harlequin Green
617	618	Harlequin Blue
619	620	All Other Miscellaneous Rares

DIVISION 6 — EXHIBITION BUDGERIGARS

Awards: *Best Exhibition Budgerigar*
In each subdivision:
Best & 2nd in each section
Best Hen in each section
Best Young in each section

There will be three subdivisions: Champion (C), Intermediate (I) and Novice (N). Birds entered in the Novice Subdivision must be bred and banded by the exhibitor. If you are a Novice breeder, put an N in front of the class number on the show-cage tag; if you are a Champion exhibitor, put a C in front of the class number, etc.

All birds entered in any subdivision in a Young class must wear a closed band of the current year. Enter your full code on the entry form. The management reserves the right to check bands.

Exhibition Budgerigars are not judged on the basis of points. The only fully satisfactory method of judging is by comparison, by placing the birds side by side and gradually eliminating the poorest, then picking out the best on the basis of straight comparison. This is the only method used by American Budgerigar Society Panel Judges and is the only method approved by the Executive Committee. Good judges are experienced breeders who instinctively place the greatest weight on those characteristics most difficult to achieve.

Section 1—Normal Greens

Cocks:		Hens:		
Old	Young	Old	Young	
101	102	103	104	Light-green
105	106	107	108	Grey-green
109	110	111	112	All Other Normal Greens

Section 2—Normal Blues

Cocks:		Hens:		
Old	Young	Old	Young	
201	202	203	204	Skyblue
205	206	207	208	Cobalt or Mauve
209	210	211	212	Grey
213	214	215	216	Violet
217	218	219	220	All Other Normal Blues (incl. Yellow-face)

Section 3—Opaline Greens

Cocks:		Hens:		
Old	Young	Old	Young	
301	302	303	304	Light-green, Dark-green or Olive-green
305	306	307	308	Grey-green
309	310	311	312	All Other Opaline Greens (incl. Yellow Grey)

Section 4—Opaline Blues

Cocks:		Hens:		
Old	Young	Old	Young	
401	402	403	404	Skyblue, Cobalt, Mauve or Violet
405	406	407	408	Grey
409	410	411	412	All Other Opaline Blues (incl. Yellow-face)

Section 5—Opaline Cinnamons

Cocks:		Hens:		
Old	Young	Old	Young	
501	502	503	504	Light-green or Dark-green
505	506	507	508	Grey-green
509	510	511	512	Grey
513	514	515	516	Skyblue
517	518	519	520	Cobalt or Violet
521	522	523	524	All Other Opaline Cinnamons (incl. Yellow-face)

Section 6—Wing Series

Cocks:		Hens:		
Old	Young	Old	Young	
601	602	603	604	Cinnamon Light-green or Dark-green
605	606	607	608	Cinnamon Grey-green or Grey
609	610	611	612	Cinnamon Blue Series
613	614	615	616	All Other Cinnamons (incl. Yellow-face)
617	618	619	620	Greywing, Green or Blue Series
621	622	623	624	All Other Greywings (incl. Yellow-face and Yellow Grey)
625	626	627	628	Clearwing, Green or Blue Series
629	630	631	632	All Other Clearwings (incl. Yellow-face)

This Normal Skyblue Budgerigar cock won Firsts in several major shows in England during the 1957–58 season.

Prize-winning White-flighted Blue Pied Budgerigar.

Section 7—Lutinos, Albinos, Fallows

Cocks:		Hens:		
Old	Young	Old	Young	
701	702	703	704	Lutino
705	706	707	708	Albino
709	710	711	712	Fallow
713	714	715	716	Opaline Fallow
717	718	719	720	All Other Red-eyed (incl. Yellow-face)

Section 8—Miscellaneous Rares

Cocks:		Hens:		
Old	Young	Old	Young	
801	802	803	804	Pied Green
805	806	807	808	Pied Blue
809	810	811	812	Opaline Pied Green (incl. Cinnamon- or Yellow-flights)
813	814	815	816	Opaline Pied Blue (incl. Cinnamon- or White-flights)
817	818	819	820	Harlequin, Green or Blue
821	822	823	824	All Other Pieds or Harlequins (incl. Yellow-flights)
825	826	827	828	All Other Miscellaneous Rares

DIVISION 7 — COCKATIELS

Awards: *Best Cockatiel*
             Best in each section

All Cockatiels should be entered in this division. However, all Normal Cockatiels are eligible for the Best South Pacific Parrot award in the Parrot division, and all mutation Cockatiels are eligible for the Best Mutation Parrot award.

Section 1—Normal Cockatiels
101   Old cocks
102   Young cocks
103   Old hens
104   Young hens

Section 2—Pied Cocktaiels
201   Cocks
202   Hens

Section 3—Lutino Cockatiels
301   Cocks
302   Hens

Section 4—Pearl Cockatiels
401   Cocks
402   Hens

Section 5—Cinnamon Cockatiels
501   Cocks
502   Hens

Section 6—All Other Varieties
601   Split Pied Cocks
602   Split Pied Hens
603   All Other Variety Cocks
604   All Other Variety Hens

Normal Cockatiel cock

DIVISION 8 — DISPLAYS

Awards: *Best Display (not eligible for Best in Show)*
                      2nd & 3rd Best Display

Section 1—All Displays
101   Canary Displays
102   Budgerigar Displays
103   Finch Displays
104   Parrot Displays
105   Dove Displays

Until presentation, the winners' trophies are usually arrayed prominently in the show hall.

Above: This view of an exhibitor's hotel room hints at some of the logistical problems involved in showing birds.

Facing page: The 1980 Boston Society for Aviculture Show. Young visitors find their attention captured by the trophy table (*above*) or by a Red-lored Amazon (*below*).

Not all shows have a division for displays, which are large cages of three or more birds variously decorated. These are judged simply for beauty. I well remember one attractive display I see every year: it is a miniature house with living room and bath. Housed within are several colorful canaries.

Those readers familiar with show catalogs undoubtedly will have noticed still more differences between the above schedule and others they have seen. The terms *old* and *young*, for example, are used throughout, even in those sections where *flighted* and *unflighted* are perhaps more conventional. This has the value of reducing confusion for novices and the interested public who pick up a show catalog.

Also, the numbering of classes is the same from division to division. In the more complex divisions (Parrots and Canaries) a four-digit number best suits the purpose: the first digit indicates the subdivision, the second the section and the last two the class. When subdivisions are not involved (the Finch division) three digits are sufficient to designate section and class. A rational and consistent numbering system for the entire schedule is helpful especially to exhibitors and show secretaries and stewards, not to mention the members of the public who attend the show. These visitors appreciate being able to easily identify an exhibit by looking it up by number in the catalog.

In many show catalogs the awards are listed separately from the classes, particularly if the kind of award is stated (trophy, rosette, certificate, etc.) and if the names of sponsors appear. In any show, awards vary among divisions and sections depending on patronage and the conventions of the fancy. Best Young, for example, is frequent in those sections that have separate classes for old and young birds. While autumn bird shows usually have *open* competition (that is, open to nonmembers of the club), there may be some awards, or even some sections, that are *closed* to nonmembers. In addition, the show may offer "special

Show cage variations. *Right:* The Lancashire, one of the largest canary breeds, requires a spacious cage. *Below:* Roller canaries are mostly exhibited in teams; this is the top song cage in a stack of four.

The 1980 B.S.A. Show. *Above:* Dr. Decoteau about to begin judging the Cockatiel Division. *Facing page, above:* Dr. Decoteau placed a Cinnamon hen First in Division, while a Normal cock was Third; both birds were exhibited by Linda Rubin. *Below:* In three sections (Normals, Pieds, and Rares) First were taken by hens.

At one cage-bird exhibition in England, the development of Budgerigar varieties from the wild form (on the left) was displayed in this fashion.

awards." One catalog, for instance, lists a "Diploma for the highest-scoring Pastel Colorbred Canary of any ground color."

Every schedule of classes includes some rules or instructions pertinent to only a particular part, such as the use of the Group System in judging American Singers or the separation of Exhibition Budgerigar entries into Champion, Intermediate and Novice subdivisions. Show-giving clubs eventually must abide by the rules and regulations of the national or international breed organization. Almost all breed shows will agree to use the standards set by these specialty associations. At times a specialty society will sponsor all classes for their breed, giving special awards.

Besides the rules relating to particular breeds, show committees publish a list of general rules in their catalogs. The following illustrate typical points:

1. The entry fee for all divisions shall be 75¢ per bird entered.

2. All birds must be registered upon entry to the show hall. A separate entry blank must be used for each division. Present the entry blank to the Cashier with entry fee, then proceed to the Division Secretary.

3. If there is a standard show cage for your breed, we suggest you use it. Show cages are designed to show your bird off to its greatest advantage. Be sure you use perches of proper size for your bird's feet.

4. All cage bottoms must be covered with seed, unless specificed otherwise in the Schedule of Classes.

5. All entries must be registered and benched before judging has started.

6. Exhibits cannot be removed before 5:00 PM. This is for the convenience of the paying public.

7. Birds are to be entered singly, one bird to a cage.

8. All exhibitors must conform to show rules and adhere to requests made by the Show Chairman.

9. No names, distinctive marks or decorations of any kind will be allowed on cages until after judging has been completed.

10. Any exhibit entered in a wrong class can be reclassified with the approval of the judge.

11. All decisions of judges are FINAL.

12. We reserve the right to refuse admittance or to remove any bird showing signs of illness.

The 1980 B.S.A. Show. *Above:* The exhibition of a White-breasted Gouldian Finch created a stir of interest among finch fanciers, many of whom had not seen this variety previously. *Facing page, above:* Budgerigar judge Don Langell assessing entries. *Below:* Some of the winners.

On the whole, in number of entries English bird shows far outstrip those held in the United States.

EXHIBITING

SHOW CAGES

Certain types of birds have standardized cages that must be utilized when exhibiting. Canaries and Budgerigars must be shown in specific show cages, but finches, parrots and softbills can be shown in any neat, suitable cage. Most parrots are exhibited in typical parrot cages. Some macaws and larger cockatoos are exhibited in large dog cages or custom-built parrot cages. Finch exhibitors often use a Canary show cage, and lovebirds are frequently exhibited in budgie show cages.

The American Cockatiel Society has developed a very nice show cage to be in full use by 1983. The outer color is black and the inside is a sky-blue color. Meanwhile, Cockatiel exhibitors can utilize any suitable cage.

Cages should be neat and clean. They must not be chipped; freshly painted cages look much better on exhibit. Clean seed is placed on the bottoms of all cages. Shavings, newspaper, soil and similar substances must never be used for exhibition. Most important of all, the bird being shown must be comfortable, with sufficient room to move about freely without damaging its tail and wing feathers. Too often a poor cage and poor presentation have kept a good bird from winning.

PREPARING FOR FINCH CLASSES

At exhibitions the finch classes of birds are frequently called "foreign" birds. The English exhibit these birds as

The 1980 B.S.A. Show. *Above* Jose di Almeida's borders won First, Second, and Third Best Young. *Facing page, above:* John Bradbury judged Border and Type Canaries. *Below:* Staging of some of the canary entries.

Show cages are designed to accommodate and present a breed to its best advantage. *Left:* Yorkshire Canary show cage. *Below:* Border Canary show cage.

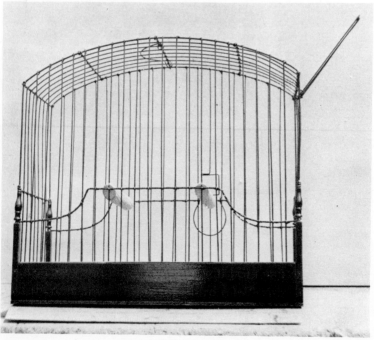

Budgerigar show cage. The regulation size of show cages simplifies staging and makes for fairness in judging.

pairs, but in the United States they are most often exhibited as single entries.

Cleanliness is of utmost importance in keeping birds in good feather. We suggest spraying finches with a fine, luke-warm mist spray. If a finch has just been through a molt, it perhaps has a few pin feathers apparent, particularly on the head. It should not be exhibited, because in strong competition an entry in this condition has no chance of winning. Also, if the bird is still molting, it could be detrimental to exhibit the entry since its health might be threatened.

It is suggested that finches be placed in their show cages at least four days before the show, or even sooner if the exhibitor has the time. Here they can become adjusted to their surroundings, well in advance of being judged. Occasionally a wild finch may fly against the cage bars and may injure or destroy its feathers. It is best to show a bird that has all tail feathers and wing feathers. A judge may take off

The 1980 B.S.A. Show. *Above:* Type Canary exhibitor John Magee holding his Best Yorkshire and Best Old Border. *Facing page, above:* Judges, officers, and members enjoying themselves at the Saturday-night banquet. *Below:* After dinner, B.S.A. president Linda Rubin (*left*) presented the Harry Ross Memorial Award to avian veterinarian Dr. Marjorie C. McMillan for her "outstanding service to New England aviculture."

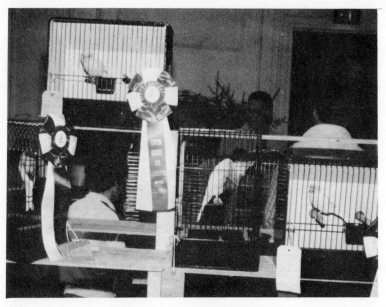

Some winning canaries at the 1980 B.S.A. Show.

points on birds with missing feathers, so one must be careful when catching finches for show. On one sad occasion I was about to catch an excellent fire finch to place into the exhibition cage. Upon catching the finch, every tail feather came out. That poor little finch of course had to remain at home.

Most often cages used for exhibiting finches are painted black on the outside and white on the inside. The bird must be unable to see its reflection on the inside white, so a matte finish is required.

One should use a seed covering on the floor of the exhibition cage. This looks neat and can be utilized for finch food during the exhibition. Shavings, sawdust, paper and other related items must not be used.

One must ensure that the perches in the cage are appropriate for the species exhibited. If the perches are too thick, the finch will have difficulty gripping them, and if the perches are too thin, the bird may not show to best advantage.

AN EXHIBITOR'S DAY AT A SHOW

Your day might begin at 4 AM, since some shows are some distance away from home. This should cause little difficulty, for if you are a fortunate planner, you will already have trained your birds in suitable show cages. By the morning of the show, each bird should be confident and calm in its temporary home.

Upon arrival at the show, it is wise to first find the entrance and make a dry run to see just where the birds should be taken for entry and examination. Many show committees require that birds be examined in a special room by a show veterinarian. Birds must also be entered properly before they can be benched in the exhibition hall.

You will have to make out the entry application, generally a different one for each division of birds, and then present it with the proper fee (either seventy-five cents or one dollar at most shows) to the show secretary. If you are entering a finch, a parrot and two canaries, then you must have three different entry forms.

Once your fees and forms are submitted and your birds have been examined, you will be given an entry card to complete for each entry. This card states the entry number, class number, section number and division number, which are all the judge will see. The card also contains space for your name and address as well as space for the species of bird exhibited. The card is designed to fold over your name and address and bird species when stapled, which, by the way, is mandatory. It can be opened only when all judging is completed. This card, or tag, is tied on the show cage of the bird entered.

Once all identification tags are attached and stapled and your cages are in place within the exhibition hall, you should check that each bird is comfortable. Fresh water must be given. All the cage floors are already covered with seed, a standard practice. Perhaps your birds each desire a

The 1978 New Hampshire Cage Bird Association Show. *Above:* The author's Scarlet Macaw was judged Best South American Parrot, then went on to Second Best in Show. *Facing page, above:* Best in Show went to a Pearl Cockatiel exhibited by Nancy Reed. *Below:* Among the 480 birds entered were these cockatoos; note that, in the absence of show-cage regulations, birds may be exhibited in any suitable cage.

piece of apple or some other tidbit. Just prior to judging, all water containers will be removed.

By this time, and after a break for a cup of coffee or whatever, you'll find it will be close to 10 AM. It is then judging time. The stewards and judges' secretaries will be bustling about with last-minute details. They will perhaps be lining up the entries by classes and by sections. It is now time to secure a good seat in the gallery from which the judging can be viewed.

You note that your parrot has five competitors. There are only three awards in that class—first, a blue ribbon; second, a red ribbon; and third, a yellow ribbon—so this means that three birds will not go home with a ribbon.

Finally, the chief steward ushers in the judge and judging begins! If you are lucky, you will have a judge who explains his placings; however, there are some judges who say little or nothing. The judge uses a stick or pointer to seek out characteristics of the birds. It seems an endless time before your class comes up before the judge, but finally the time comes—if you ever had butterflies, they were never worse than now. After what seems to be an endless period, the judge finally marks three of the tickets. Your bird made it! But in which category; first, second or third? Yours is tops today and, as the steward applies the blue ribbon on your cage, you scream with delight! The steward next applies the red and then the yellow ribbons on the appropriate cages. Your bird will be eligible for later judging in section and division, having won the class.

By midafternoon, after much fun watching various judging feats, the top finals have been judged, and the anxiety is over. Now you have time for more fun (after you water your birds again). Raffles, door prizes and looking over exhibits of bird foods, cages and other supplies will keep you quite busy. Then there is always time for "bird talk" with other novices, professionals, judges and show chairmen. Five o'clock is upon you before you know it, and a grand

Budgerigar judge Irene Evans employs a pencil to test the birds' deportment.

exodus of birds is evident. Many exhibitors go home, but some stay for the banquet put on by the show committee. Many top awards are presented at this annual banquet. Other awards, trophies and special awards are presented in the exhibition hall around 4 PM, prior to release of the birds.

Finally the long journey home begins, but it was worth it! You'll be sure to try it again!

HOW TO WIN (AND LOSE) AT THE SHOWS GRACEFULLY

Preparing and caging birds to win on the show circuit is not too difficult. In fact, with some form of ingenuity and much devotion, it is quite easy. You should study the various classes available at the shows that interest you. It will help if you can refer to a copy of last year's catalog to study the classes offered. You should watch your birds.

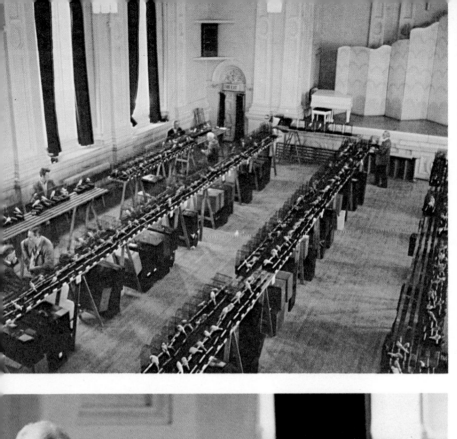

At the 1980 Open Show held by the York-
shire Canary Club of England, over 600
Yorkshires were placed on exhibit. *Above:*
The perpetual trophy awarded for Best in
Show; the "shields" around the base bear
the names of previous winning exhibitors.
Facing page, above: The show was held in
Victoria Hall in Saltaire, West Yorkshire.
Below: The large number of entries required
the services of five judges; here C.J.
Roberts judges his classes.

This Border Canary breeder has put his best birds into show cages. Steadiness in the show cage is one of the factors to consider when deciding which birds to exhibit.

Those in best condition are the ones to train in show cages for eventual competition. You should hope for much competition. It is more meaningful to win a third prize in a class of twelve than a first with no competition.

The human qualities required for the successful breeding and exhibiting of birds are not much different from those needed to succeed in any other venture, whether it is a simple pastime, a serious hobby, running a business or sailing a boat. The most successful will be those people who have a keen interest in what they are doing. Included are those who do their homework or have a special aptitude for breeding birds. Likewise, those who have tenacity and take what they are doing seriously will reap the rewards of showing birds.

Such people are often fanatics. In the bird hobby, they allow the projects to take up a large part of their lives, almost to the exclusion of everything else. They put a great deal into the hobby, and they expect—and most often get—a great deal out of it, including winning in many competitions. It is these very people who are collectively responsible for pushing and breeding birds to perfection. These are the producers of those beautiful mutations; it is doubtful whether the present-day varieties would be half as good as they are had there not been past and present generations of dedicated bird fanciers.

Ultimately, the people who derive the greatest satisfaction from their hobby of breeding and exhibiting birds are those who can gauge the rewards they reap against the effort they put into it and feel satisfied. The point I wish to make is that if, with much time and emphasis in advance, you practice showing your birds and if your birds fit the standards published in various breed organizations' literature, you will have lots of fun and plenty of satisfaction with your results.

Border Canaries, White and Blue. In the opinion of many Border fanciers, these drawings by R.A. Vowles, though done some years ago, still depict the ideal Border.

94

Ideal Border Canaries, Cinnamon and Green.

This trophy collection shows that the exhibitor was able to win with his birds over a considerable period of time.

JUDGING IN THEORY AND PRACTICE

WHAT THE JUDGE MUST LOOK FOR

Once a person has bred and exhibited cagebirds for an extended period of time, he or she may wish to specialize as a judge of a certain kind of bird. There are Canary specialists, budgie specialists, finch specialists, parrot specialists and exotic specialists ("exotics" covers finches, parrots and softbills). Occasional all-arounders are eligible to judge everything—but there are obviously very few of these.

When confronted with a class of birds, a judge must consider two major things. First, he must consider each bird in competition against every other in the same class. Second, he must weigh each bird against the ideal. Often club associations have a declared standard of perfection which sets up the qualifications of the ideal bird.

The major categories for the judge to review include:

Conformation: This—also called *type* or *form*—covers the general body build of the exhibition bird. Does the bird conform to the ideal? Is the bird too chesty? Is the tail of sufficient length? Does the bird have a smooth topline? Does the bird appear to have a pinched tail?

Deportment: Stance can be included in *deportment.* How well does the bird perch? Does it fly about needlessly? If so, is it a bird that *must* fly constantly? Certain finches must keep moving for best display, and there are waxbills that fly

Above: Red-lored Amazon in a slouching stance. A judge prefers to see a bird erect, at a sixty-degree angle.

Facing page: A Plain-colored Amazon displaying ideal stance on the perch. The slate tone over green is ideal for this species.

from perch to perch, moving their tails from side to side constantly. The judge must know these exceptional characteristics.

Color: Is the color dull and drab? Is there smuttiness in the depth of color? Obviously the judge must be familiar with the normal coloration of the exhibition bird. In breeds that have many new varieties, the judge must keep himself familiar with constant changes.

Condition: Is the bird too thin? Are the feathers damaged? Are there bald spots on the bird? Is the bird too fat? These are some of the questions a judge can ponder.

Presentation: Cage condition is another way of speaking of presentation. Many judges will count points off birds exhibited in poorly built cages. Cages that are too small or even cages that are poorly lighted have a bearing on the judge's thoughts.

Scoring an American Budgerigar entry: judge Linda Rubin at the 1980 B.S.A. Show.

Borders being trained to the show cage. Attaching the show cage to the box cage allows the birds to become accustomed to the strange enclosure as they will.

Rarity of Species: Some judges take into consideration the rareness of a species of bird, giving more points to the rare bird. Other judges do not take this into consideration.

In total, the judge takes many things into consideration. However, he usually judges by comparison, bird to bird: which one fits the ideal to a greater degree?

CONFORMATION

Most important in the eyes of the judge is the conformation of the bird or birds exhibited. This characteristic, more so than any other, involves the comparison of the bird being judged with the ideal. The judge must determine how well the individual bird approaches that ideal.

The Cockatiel, for instance, should be a long, slender, classy bird with about a two-and-a-half-inch crest. A judge should not consider as best a dumpy, fat little bird of short stature with a one-inch crest.

Paradise Whydahs. The cock shown above
has poor stance and crooked legs, while
the hen on the facing page exhibits
poor feather condition.

Paradise Whydah hen.

The entire bird must appear well balanced. The judge must look at the topline of the bird. Starting at the top of the head and going along the topline to the tip of the tail, there should be an unbroken continuity. The head should not suddenly become flattened; rather it should gently curve to the rear. I often notice abrupt changes at the point of the neck, as if between the head and back a V-shape occurs. Where the neck is prominent, the line should flow to the top of the back without abrupt changes. The back should generally be straight through to the base of the tail. Too frequently one notes a "pinched" tail which causes the tail to elevate too vertically. This condition is frequently seen in Society, or Bengalese, Finches. No hollow areas should appear, particularly at the shoulders.

In certain parrots there is a crossing of the wings, but in most birds the wings should be set correctly and tightly to the body with no crossing over. The judge must be familiar with those species having normally crossed wings.

One looks at the chin and upper throat on a line with the breast, abdomen and undertail coverts. The judge should look for bulging breasts, pinched abdomens, prominent thighs, pinched tails and oversized vents. I once judged a nice class of Zebra Finches. Fortunately most were of good conformation, but two were abnormally large in the breast and their conformation was off considerably. In particular, the judge must have excellent knowledge of the build of the species that is being judged. A good example of this involves the tails of waxbills. These birds have the ability to fan out their tails, which is a desirable characteristic.

Once a depiction of the ideal is released by a specialty organization (such as the Cockatiel society has done), one should study and restudy it. An aviculturist may wish to select breeders that more closely conform to this standard. Their progeny may fit the ideal to a greater degree and thereby impress not only the judges but you, the breeder, as well.

Blue-capped Waxbill. The too heavy breast is a deficiency in conformation, as is the V-neck.

Above: Of these three cockatoos, the Lesser Sulphur-crested in the center is in the best condition. The Lesser in profile shows tail fringing, while the third bird, a Red-vented, has poor feather condition overall.

Facing page: Another example of deficient feather condition. Note the bare areas beneath the wing of this Mealy Amazon.

107

DEPORTMENT

One of the most important aspects of exhibiting and judging involves the deportment, or action, of the bird exhibited, steadiness and stance contributing to this deportment. Although more points are given to condition and conformation, quite often a bird with poor deportment will be put down by the judge.

Deportment involves much training by the exhibitor. It is wise to place the bird to be exhibited into the show cage weeks before the first show of the season. Leave the bird in the exhibition cage for about two days. Approach the cage frequently and allow many strangers to approach the cage, since the judge will be a stranger. Try to see if the bird will stay perfectly still on the perch. Keep coaxing the bird to climb onto the perch every time it descends to the cage bottom. In due time it will find the perch to be the most comfortable place in the cage. Repeat this procedure again periodically.

If a bird shows poor deportment, it may flutter all about the exhibition cage; this is one of the most serious displays of poor deportment. If the show bird prefers to sit hunched up in a corner on the cage floor even after the judge attempts to get it back on the perch, then this is poor deportment too. Only much training can avert poor deportment. Four days before the show, place your bird in the show cage for the remainder of pre-show time.

I well recall one judging assignment in which 22 outstanding lovebirds were presented in various classes. One particularly fabulous-looking lovebird simply refused to budge from the floor of the cage, so I could not fully assess all of its qualities. Unfortunately, I had to put it down in comparison to other lovebirds with good steadiness and stance. Later the owner (in agreement with my placings) indicated that a nudge or two after opening the cage door eventually put the lazy lovebird on the perch. Of course, a judge cannot open the cage door to do this! Once on the

Training in deportment. This Budgerigar breeder is accustoming one of his birds to an experience it may have while being judged.

Comparison for feather condition. The Red-rump cock above shows poor feather condition.

Red-rump cock. The feather condition of this specimen is evidently better than that of the bird shown on the facing page.

perch, after much nudging, it appeared that the lovebird was outstanding in condition as well as conformation, and it may well have won. It is worth mentioning that when I approached the cage again, the lovebird immediately hopped down.

COLOR

Too often I have seen judges swayed by color alone. Of course, it is inviting to see a colorful, bright-red parrot or a rainbow-colored Gouldian Finch. A judge must be fair and consider all points: deportment, conformation and condition as well as color and other less important features. A gray Cockatiel may have clear, rich lines and indeed be superior to the dull red of a rosella.

Patterning should be as distinct as possible. With the exception of pieds, no odd colors, shading or smuttiness should occur in areas that must be pure. Smuttiness is seen most often in macaws and amazons as black, smudgy pigment scattered within green feathers. Another defect in coloring occasionally seen is one color streaking into another. A good example is the orange ear patch of the Cockatiel. This patch must be circular, clear and rich, but sometimes it sweeps down into an irregular patch—a defect.

Color is important, although condition and conformation appear more heavily weighted on a scale from 1 to 100. Keep in mind that there is an overlap between color and condition. For example, the beak of the Zebra Finch might be very pale due to poor condition; thus the color of the beak has double meaning to the judge. There is a blending of all characteristics considered by the judge.

CONDITION

When a judge is given a particular class of birds to judge, it is always a good idea for him to walk around looking at each bird in the class first. During this quick review, the

Black-cheeked Waxbill. Overall poor condition is emphasized by the fluffed feathers.

judge can secure a count of the number of birds and at the same time also note the general condition of each bird. From the beak to the tip of the tail, a quick view gives the judge a good first impression of each bird.

Then he can concentrate on studying each bird critically for an extended period. This is when one must look for dirty and soiled feathers, plumage that appears too loose to the point of fluffed feathering, or frayed tail or wing feathers. The most often noted condition which degrades a bird involves the tail. One sees frayed and damaged tails much too often. This discredits the bird in large classes, almost always precluding that first-prize blue ribbon.

A small or cramped cage or the last-minute preparation of the bird for show often creates the damaged tail. Ideally, the exhibitor should place the exhibition bird in its show cage (which must be sizeable and comfortable) several days

Above: Diamond Sparrow. Exhibiting vivid color, it also appears in good condition, except for a defect in the conformation of the beak.

Facing page: White-fronted Amazon in fine condition, its color strong and vivid.

Quail Finch in poor condition; note the loose feathering, faulty growth of the upper mandible, and untrimmed claws.

before the exhibition. This calms the bird. If it still damages its tail or wing feathers, through manipulation and finely spraying the feathers, the bird may be more presentable by show time.

It is surprising how a little water in a fine spray will clean these birds. One should use a fine, lukewarm mist spray two to three times per day on show birds. Fresh rain water appears to do a better job than tap water. Eventually, the feathers simply shine.

Wing feathers are sometimes battered on both primary and secondary endings. This too downgrades a bird on condition. Judges look for close, smooth, glistening plumage. If the plumage is dry and rough, it is detected and marked accordingly in the judge's notes.

The beak must appear normal, not overgrown or crossed. Claws are frequently allowed to become much too long. I once judged a nice class of finches which included a fine, well-balanced Bullfinch. The claws were so extremely long that in the final judging for Best Finch it had to be put down because of those very long claws. With just a little time and a pair of scissors or claw snippers, this bird could have been sitting beside the Best Finch trophy.

Legs and feet in general should be smooth. A missing toe or a crooked toe is not too important unless the competition is keen.

DISQUALIFICATIONS AND DEFECTS

Birds and small mammals have been exhibited for centuries in various cultures. The English are avid fanciers of all types of birds and mammals, having contributed much to their development and having had exhibitions on record for two centuries. Since, with the exception of Canaries and Budgerigars, the exhibition of birds is still in its infancy, we in the field of aviculture have much to learn. There will be very much debate about many activities involving the exhibition of all birds, particularly the exotics. Often, cer-

Above: Military Macaws. The one is in good condition, while the other is missing its tail.

Facing page: Lutino Cockatiel. A bald spot occurring behind the crest is a defect frequently found in this variety.

tain show restrictions are initiated with little thought. The various organizations must constantly review their standards to ensure that thought and common sense as well as good judgement prevail.

In the exhibition of birds and mammals, any *disqualifications* listed are conditions or deficiencies that are passed from parent to progeny. These inherited conditions are more detrimental to a lineage of birds than any injury or defect in an individual.

It is disturbing to note that some associations call a missing toe or toenail a disqualification; classically, it should be characterized as a *defect*, not a disqualification. There are many outstanding birds that will produce top young stock that may be missing a claw due to some larger bird, perhaps a macaw, taking advantage of that toe through a fence or otherwise. Why penalize the bird that much? True, it is a defect and should keep the bird from reaching Best in Show. However, let us keep in mind why we really exhibit. Are we not potentially showing off the stock that will produce progeny with improvements? How often have you attended a show and noted a "toenail judge?" The bird can be a perfect specimen with good stance, excellent feather and condition, and good conformation, but because one toenail is missing, the judge throws it out.

We should look at each bird in its entirety and also learn from the long experience of other exhibitors of birds and mammals. It is never too late for standards to be changed. In fact, standards constantly need changes. As exhibitors and judges we must consider missing toenails and the like as defects, not disqualifications.

RARITY OF SPECIES

There are many judges of exotic birds who consider rarity of species as part of the judging system. Certain judges give as many as 15 points out of a possible 100 for rarity of bird. Others give 10 points, while some give only 5 points.

Hawk-headed Parrot in excellent condition, characteristically erecting its nape feathers at will.

Approximately 70% of the judges of exotics do not utilize rarity as part of their system of judging.

Personally, I feel that each bird must be exhibited on its own merits of conformation, condition and deportment. Why should a run-of-the-mill but rare Hawk-headed Parrot be placed over an outstanding Normal Cockatiel hen?

On many occasions various exhibitors have become temperamental when another bird defeated their bird because rarity made the difference. They eventually get over the difficulty, coming back the next year stronger than ever, but now hoping that a different judge pays little attention to rarity.

Rare one year, not the next. While this occurred in the case of Goffin's Cockatoo (*left*), it is extremely unlikely in the case of the Palm (*below*).

Of the rosellas, the Pale-headed is one of the more common in captivity.

We feel it is more important for a judge to examine a bird with the premise that the training process of steadying the bird for deportment is more important than rarity. There are some exotic birds (many of them rare) that always seem to be ready for the judge. Good examples are the rosellas, the Gouldian Finch and the Bullfinch. Many cockatoos are elegant, needing little grooming and training. But a typical conure trait is to damage wing and tail feathers from extreme activity. Apparently the rosella is a more docile bird; consequently, there are generally no messed up and damaged wing and tail feathers. Therefore, an extremely rare conure should not get those 10 or 15 extra points simply for rarity. I have seen this happen at a show too often. Such a point system brings up a poor-looking rare conure to defeat a good-looking but more common rosella.

Many debates will continue on whether rarity should be a factor in judging. Some birds may be rare today and common next year. This happened with Goffin's Cockatoo: in 1973 this was an extremely rare bird in the United States, but by 1977 it was very common.

HYBRIDS AND MULES

While certain Canary enthusiasts marvel at the hybridization of certain Canaries and finches with the objective of producing a mule or hybrid for better song or better color, other aviculturists look down on the crossing of pure species of birds simply for no particular purpose. (Some hybrids have the ability to reproduce, while those that are called mules cannot.) There is currently a move to prevent the hybridization of the exotic species. Certain show-giving clubs are also deleting classes for all mules and hybrids. The American Federation of Aviculture has gone on record against hybridization.

For myself, I believe it is desirable to maintain pure breeding stock. With all the new varieties being developed,

A fledgling begs from its Goldfinch parent, while Bullfinch juveniles remain impassive. When compatible species such as these are housed communally, hybridization may occur.

particularly in the world of lovebirds and Cockatiels, there is no reason why we should encourage hybrids in our exhibitions.

Crossbreeding through the ages has proved that results are always unpredictable and never constant. Heaven knows, the judges have enough pure breeds to study and comprehend, without being faced with an unusual specimen that makes them wonder, "What new species is this?"

There is one good reason for crossbreeding. This involves the development of a new breed. By generation after generation of crossbreeding and selectivity, including the use of outcrossing again and again, one may eventually develop a new variety. However, all progeny or offspring must have the same characteristics in color, size and conformation; they must breed true each time. This has been accomplished in fancy poultry, with many new breeds developed over the years.

Interspecific hybrids.
Left: Peach-faced x
Masked Lovebirds.
Below: Goldfinch x
Linnet.

"Shamrock" Macaw, the offspring of Scarlet x Military. Individuals of this cross vary considerably in coloration, so that often they are not recognized as what they are.

Another area of notable hybridizing is with the finch-canary crosses. Here a group of bird enthusiasts have mastered the art of hybridizing various finches and canaries for better singing characteristics, as well as for new color combinations. For color, the finch most often considered is the Hooded Siskin of South America, crossing males of the Siskin with female Canaries. It is thought that many of today's colorbred Canaries, some of which are deep red and deep orange, carry the blood of this siskin.

Another famous cross involves the European Goldfinch, quite a singer in its own way, and the Canary. It is interesting to note that head color, though lighter and not too distinct, is passed from the Goldfinch to the young hybrids.

The Bullfinch is frequently crossed with the Canary for the production of a hybrid with virtually no neck. These are attractive hybrids. Greenfinches, Redpolls and even American Goldfinches have been used in crosses. Chaffinches and hawfinches are also used occasionally.

The English are ardent fanciers of hybrids and mules, and they frequently have hundreds of entries in these classes. On two occasions within the last few years the Best in Show at the National Show in London was a hybrid or mule. This is indeed impressive for those exhibitors that fancy their success in breeding this special kind of bird. In the late fall of 1980 at the Hempstead, New York exhibition, I judged a fine group of mules and hybrids that were classed in the same division as the finches and softbills. Out of 55 birds entered, I placed a fabulous classy and showy mule that was of non-European origin. In the United States it is seldom that a mule or hybrid will go all the way to Best of the entire division.

EXHIBITING SINGLES OR PAIRS

The English are noted for their fine exhibitions. In many of their finch classes, particularly Zebras and Societies, they seem to prefer the exhibition of pairs over single birds.

In the United States, although some show-giving organizations do allow for pair classes, most prefer single-bird entries. There are also those judges who prefer true male-and-female pairs, while others will accept pairs regardless of sex.

Most associations and judges realize that if pair classes are offered, then both birds must be practically identical. It is very difficult to get a pair of birds into identical condition and often tougher to select for identical color markings, particularly in Society Finches. It is even more difficult to keep a pair of birds in top condition throughout the entire show season.

One incident I remember well involved a show in which a pair of Penguin Zebras were exhibited against several single Penguin Zebras. The hen was in horrible condition, but the male was one of the most superb Zebras I had seen. Since a judge must judge both as one entry, I had to put this entry down with no award. The male alone could have gone all the way to Best Zebra and possibly Best Finch.

I have even had parrots before me for judging in which two different species were in the same cage. This is (and should be) forbidden in all shows.

The point to be made is that both birds in a cage exhibit must conform to their utmost, or one can pull the other down to defeat. If classes are offered for "pairs only," that is a fairer approach to judging them. However, they are still affected in the finals of judging. The English have had pairs that ended up Best in Show. I well recall a pair of White Zebra Finches that did this in the last decade, and they were indeed splendid. In the United States, however, there are very few pairs that ever end up on top. The current trend here is to exhibit single birds in individual show cages.

As with canaries, finch exhibiting in the U.S. follows many of the guidelines set down by English specialist societies—in this instance, the National Bengalese Fanciers' Association and the Zebra Finch Society. If finches are to be exhibited in pairs, the two birds must be appropriately matched. With Society Finches, the requirement that they be similarly marked can present difficulties if Chocolate-and-Whites or Fawn-and-Whites are being shown; varieties such as Chocolate Selfs (*above*) present little problem in this respect. With Zebra Finches, similar markings are necessary if both birds are the same sex or if they are Pieds. In cock-and-hen pairs, both must be of the same variety, like the Chestnut-flanked Whites shown on the facing page. These birds illustrate one of the difficulties in breeding this variety: maintaining the intensity of the markings.

Standards that apply to many different groups of birds must be formulated in general terms; hence, they tend to sound the same. However, the judge applies the standard with the birds in question in mind. Different values come into play when assessing the deportment of a Blue-fronted Amazon (*above*), as compared to that of a glossy-starling.

SOME BIRD
STANDARDS

BIRDS THAT HAVE STANDARDS

As we progress toward greater expertise in exhibiting birds in the United States, we note an ever-increasing interest in the development of standards for judging. It is indeed difficult for a judge to select winners without a standard. Certainly, as will be evident by examples later in this book, the Canaries have well-organized judging standards. A standard, of course, is only as good as the judge utilizing it. In the finch series, both the Bengalese and the Zebra Finches have set standards. Within the past three years, the American Cockatiel Society has developed a standard for Cockatiels, both normals and mutations. The African Lovebird Society has likewise developed a lovebird standard, again for both the normals and the mutations.

Standards are important and helpful to the judge. However, standards are to be improved upon and must be based on selective breeding toward the ideal. It has been known to happen in the development of some standards that certain strong individuals have effected a standard to suit their strain of a breed. In fact, in one show dog the standard in the 1950's required the breed to be comparatively low on legs with a short, cobby body. But in viewing the standard of 1980, we see a drastic change to a long-legged creature with an abnormally long body that would not have won tenth prize, much less second or third,

Above: This Eastern Rosella should not be exhibited until its feather condition has improved.

Facing page: A Yellow-collared Macaw in good condition, with great color.

A general parrot standard must encompass species as diverse as the Yellow-crowned Amazon (*above*) and the Yellow-fronted Parakeet (Kakariki) shown on the facing page. In each instance the judge must evaluate according to those factors which characterize the species.

Yellow-fronted Parakeet (Kakariki)

in 1950. Only the breeders who play an active part in the breed organization can make sure that changes in the standard are for the better and not for the worse. Unfortunately, it is still possible for a dominant breeder to swing the vote in favor of his strain.

Following are some general standards for various kinds of birds. Though none are exact copies of specialty-society standards, they are presented as guides for both exhibitors and judges.

A PARROT STANDARD

A general parrot standard which is utilized by some judges, though very broad, is outlined here.

Condition—30 points. This considers cleanliness, condition of the feathers, wing and tail damage, claw length and appearance, as well as missing feathers.

If the Blue-crowned Conure above and the Pesquet's Parrot on the facing page were competing against one another, it appears that both would score well with respect to Condition, as well as Color and Markings. It's likely that Deportment will be the deciding category.

Certainly the condition of this Malabar Parakeet is superior; assessment of its conformation, however, demands that the judge possess the wide experience that includes acquaintance with this uncommon species.

Conformation—30 points. This is most general for the parrots, but with much experience as a breeder of parrots the judge will have the knowledge concerning the ideal conformation of a macaw, an amazon, a cockatoo or a conure.

Deportment—20 points. A parrot that flutters all over the cage will lose many points under deportment. Likewise, one that sits on the cage floor also loses much for deportment.

Color and Markings—20 points. In parrots, color and markings can be important. Brightness of color can be overwhelming when comparing individuals of the same species; the judge must take into consideration that there are sometimes sex differences. In South American or Afro-Asian species we sometimes see a "smuttiness," irregular black markings within the green feathers, mostly on the back. This is noted as a fault.

COCK
(NORMAL)

HEN
(OPALINE)

E. H. Hart's drawings of ideal Budgerigars—a Normal cock and an Opaline hen—emphasize conformation and markings.

141

Both the Panama Amazon above and the
Light-green Budgerigar on the facing page
are in suitable condition for showing. But,
at the present time, the Budgerigar faces
explicit challenges in competition that the
Panama does not. While the Budgerigar will
be faulted at once for its missing spots,
will there be a day when size will be a
deciding factor in judging Panamas?

Lutino Cockatiel. Individuals of this variety exhibit considerable variation in color. Depending on the amount of carotenoid pigment present, they may range from an almost pure white to a buttercup yellow.

Normal Cockatiel cock in one of the commercially available cages designed for Cockatiels. Beginning with the 1983 season, shows patronized by the American Cockatiel Society will urge use of a regulation show cage.

A COCKATIEL STANDARD

A typical Cockatiel standard employs the usual categories, i.e., conformation, condition, color, deportment and presentation. Most important is the necessity that the judge must look at all the birds in a class and judge by comparison.

When one looks at *conformation*, it is necessary to visualize a long, not too bulky bird. A Cockatiel should measure about 17–17½ inches from the top of the crest to

Finsch's Conure. This bird will receive high scores for both condition and color.

Slender-billed Conure. Another example of good condition and color. However, the characteristic behavior of the Slender-bill is different from that of more familiar conures, and this must be taken into consideration by the judge.

the tip of the tail. The wings must be straight and not cross-ed at the tips, as is frequently seen. The tail should be long, straight and fully connected to the body in a streamlined effect. The neck is long, carrying a smooth topline to the back and tail. A full chest is desirable, yet it must not be so bulky that the bird is top-heavy. Often when one sees an overly large chest, one also detects a humped back.

When discussing *condition*, one can visualize the care of the Cockatiel. One of the more noticeable problems of condition involves fringing of the tail and wings, as well as loss of tail and wing feathers. Often one detects dirty tail feathers, particularly on white Cockatiels. Pin feathers also detract from condition; many times pin feathers are present on the head and neck.

Deportment can make or break a Cockatiel during an exhibition. If a judge comes upon a cage with a fluttery, frightened bird, he is likely to walk away from it. The gentle Cockatiel must sit on the perch erectly, yet docilely, at a 70–75-degree angle. Often when judging Cockatiels, I find it difficult to persuade certain Cockatiels to get up off the floor to perch. At other times, with just a tap on the cage with the judging stick, the Cockatiel will climb onto the perch as if to say, "There, judge me."

Color is also taken into consideration by the Cockatiel judge. Feeding plays a large part in the color of the bird. One must look for vivid coloring. Emphasis is placed on quality and uniformity of color, except that in Pied Cockatiels one must look for symmetrical markings. Since Pearl Cockatiel males often lack most pearl characteristics as far as color is concerned, one must admit that those beautifully pearled females may often get the edge.

Presentation of the Cockatiel, though important, is good for only about 5% of the total points. Presentation measures the cage size in relation to the bird. A judge may ask himself, "Does the tail of this Cockatiel drag against the cage wall? Is the cage large enough for the bird shown?"

Normal Cockatiel hen and cock—so called because their coloration is that normally found in the wild.

Cockatiel varieties. *Above:* Fallows, at present few in number, are not often seen on the show bench. *Facing page:* Pearl Cockatiel cock. In males pearling occurs only in juvenal plumage, which may be the reason for the sparse crest of this specimen.

Peach-faced Lovebird. This specimen shows some of the tail fringing often found on caged parrots, a result of their proclivity for climbing.

A LOVEBIRD STANDARD

Many attempts at formulation take place in the early days of new associations in order to eventually produce a single standard that most breeders, exhibitors and club members will finally approve. The lovebird fanciers formed themselves into an association in the late 1970's and progressed with much success. Several different ideas were developed to initiate a lovebird standard. By the time of their first national show, in which about 130 birds were exhibited (far surpassing any other exotic bird in numbers at any single show anywhere), they had a standard.

I prefer to set forth some typical criteria used by many judges of lovebirds, including myself. It is always best to set up a point scale to guide the judge.

Condition—20 points
Deportment—20 points
Color—15 points
Size and shape—20 points
Head—10 points
Wing carriage and tail—10 points
Legs and feet—5 points
　　Total—100 points

As in all birds, *condition* is very important. Are any wing or tail feathers damaged? Is the bird dirty? Are there any bald spots? Is the lovebird too fat or too thin? I strongly believe that condition is the most important aspect of judging.

Size and *shape* are also important. The head of a lovebird must be full and round, with well-centered eyes that are clear and bright. It should have a wide neck which appears full, with a contour from head to body that appears uninterrupted. The breast must be deep and well rounded as well as quite broad. There should be no hollow area in the backline; this line must be straight. Overall, a well-conformed lovebird that is not too fat or too thin will naturally appear

Peach-faced Lovebird varieties. *Above:* Normal and Pied Light Green. *Facing page:* Dutch Blue.

to the eye of the experienced judge as just the right bird.

Shape does to some extent take in head, wings, tail and feet, although more stress is placed on these items separately.

Deportment is also very important when judging the lovebird (as well as all birds). This can be learned by the lovebird, given proper patience and excellent teaching by the exhibitor. Nothing bothers a judge more than a lovebird that constantly remains on the bottom of the show cage or flutters when he approaches the cage. The exhibitor can prevent this by placing the show lovebird in the exhibition cage weeks before the show to ensure that the lovebird becomes adjusted to its surroundings. Repeat this procedure several times. Approach the exhibition cage in different ways. Allow strangers to approach the bird. Utilize a pointer such as the judge may use, but do not "poke" the bird with it. Soon you will have a bird with excellent deportment, and you will have earned that twenty points.

Color becomes increasingly important since there is now a rapidly growing list of various color varieties. A good judge must not favor a color, whether it be normal green, yellow or blue. In fact, the judge must look at all colors for brightness and markings. Dullness of color will definitely be detrimental. A Normal Peach-faced with a small, irregular pied marking does not succeed as a well-marked Normal and also will not win as a Pied. Another fault occurs when the facial peach color in a Normal is dull. Much judging for color must be accomplished by comparison.

Head structure is important in a lovebird. It must be full and round. The beak cannot be overgrown.

There should be no crossing of *wings*, which should be held closely and neatly in line with the body contour. There should be no drooping of primary or secondary feathers. The *tail* must be clean and neat both below and on the top side. It must also be held in line with the rest of the body.

Fischer's Lovebirds. Always one of the more fancied love-bird species, increasing numbers of Fischer's result in more appearances at shows.

Legs and *feet* must be straight and strong, with a firm grip on the perch (which should not be oversized). Missing claws must be considered as a defect, but keep in mind the total points. You cannot lose more than five points using my system.

It is indeed a treat for exhibitor and judge alike to have a large series of classes of lovebirds. To win in a large group is a great thrill for the exhibitor. The judge should love a large class—it is an honor to him.

Two rarely exhibited parrots. *Above:* Few
Cuban Amazons are held in captivity in the
U.S.; emphasis on breeding makes it unlikely that any will be placed on exhibit. *Facing
page:* Until recently, the Painted Conure
has seldom been imported.

Detail of the spangling on a Lizard Canary.

A STANDARD FOR THE LIZARD CANARY

Since this book is concerned mainly with the exhibition of the exotic and hookbill birds, we will merely mention a few typical Canary standards. The standard for the Lizard Canary is of interest. All standards vary; some are expressed with many words, some are glorious and others are short and to the point. A typical Lizard standard is follows:

Spangles—For regularity and distinctness—25 points
Feather quality—For tightness and silkiness—15 points
Ground color—For depth and evenness—10 points
Breast—For extent and regularity of rows—10 points
Wings and tail—For neatness and darkness—10 points
Cap—For neatness and shape—10 points
Covert feathers—For darkness and lacing—5 points
Eyelash—For regularity and clarity—5 points
Bill, legs and feet—For darkness—5 points
Steadiness and staging—5 points
    Total—100 points

The Lizard Canary is one of the long-established canary breeds. Usually—and properly—scheduled among the Type canaries, the Lizard is unusual in that so much of the point scale is devoted to its markings.

Above: At the 1981 N.I.R.O.C. Show, exhibitor Donald Perez won First in Division with a Yorkshire and Second with a Gloster Consort.

Facing page: Yorkshire Canary cock, which has been colorfed for showing.

Gloster Corona. The crest here will be faulted because it is poorly defined and appears tufted.

Norwich show cage. Because of the large size of this breed, the perches are placed quite close together.

A STANDARD FOR THE GLOSTER CANARY

About 1925 in England, a diminuitive breed of Canary was developed in Gloucestershire. It was named the "Gloster." There are two major varieties of Glosters: one has a crest and is called a Corona; the other has no crest and is called a Consort. A typical standard for Gloster Canaries follows:

Corona—The crest should be neat, with a regular, unbroken round shape, ensuring that the eyes are discernible; the center must be definite and notable—20 points.

Consort—The head is broad and round at all points, with a good rise over the center of the skull—25 points; eyebrow must be heavy, showing the brow—5 points.

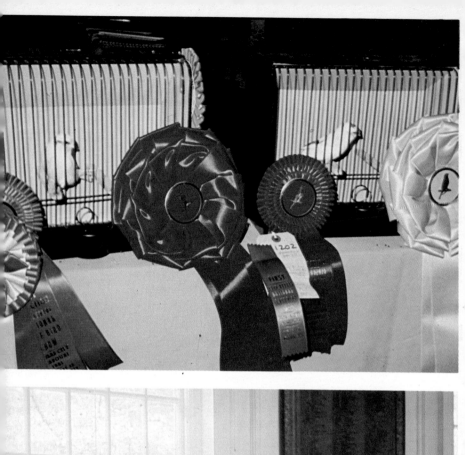

Above: At the 1981 N.I.R.O.C. Show, judge Val Clear awarded Best Foreign Bird to a Hooded Siskin exhibited by Pat Demko. The Best Hookbill was Pearl Stanley's Blue-crowned Hanging-Parrot, already Best in Division at both the 1980 and 1981 National Shows.

Facing page, above: In the Gloster Division at the 1981 National, judge John Bradbury placed Consorts exhibited by Mark Whiteaker in Second place and by Regina Buchino in Third.

Facing page, below: Judge Joe Marino appraising a Gloster entry at the 1980 B.S.A. Show.

167

Gloster Canaries, Corona and Consort.

Body—The back must be well filled, with wings set close to the body; the neck must be full, with the chest nicely rounded without any prominence—20 points.

Tail—Must be closely folded and carried well—5 points.

Plumage—Feathers close and firm, showing good quality and natural color—15 points.

Carriage—Alert and quick, with lively movements—10 points.

Legs and feet—Medium length, with no blemishes—5 points.

Size—Diminuitive—15 points.

Condition—Must be healthy and clean—10 points.

Total—100 points.

An example of the perpetual trophies awarded in England. In the United States, individual trophies are favored.

The 1980 B.S.A. Show. *Above:* Best in Show went to a Budgerigar exhibited by Dan Gallo, breeder and American Budgerigar Society Panel Judge. A Cockatiel shown by Linda Rubin was Second Best, and a Gloster Canary from Last Aviary came in Third. *Facing page:* Dan Gallo holding his winning entry, and with the plaques and trophies he won at the show.

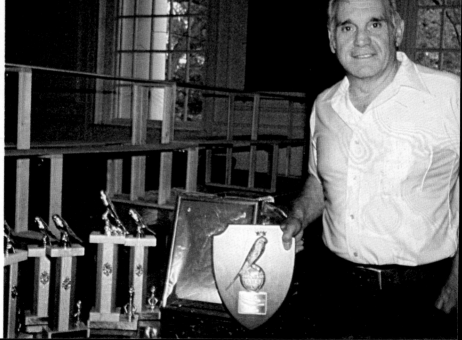

A crest-producing mutation appeared in canaries centuries ago; however, the body type of present-day Crests (*above*) and Crestbreds (*facing page*) derives from Norwich stock, and as such this breed is mostly the product of a century of selective breeding.

Crestbred Canary

A STANDARD FOR CRESTED
AND CREST-BRED CANARIES

Other than the crested Glosters, there are few other crested Canaries that have been standardized. *Size* and *formation* of the crest should be the first things to be considered. A crest cannot be too large; it should consist of an abundance of broad, long, veiny feathers evenly radiating from a small center, well over the eyes and beak. A good crest may be flat if well filled-in at the back and without splits, but a drooping and weeping crest will have preference.

Type in the crested breeds is of next importance. The body shape should resemble the Bullfinch, with a broad back, a nicely arched, full and well-circled chest, tail short and narrow, and wings neither extending beyond the root of the tail nor crossed at the tips, but fitting close to the body. The neck should be full as well as short; the bird must stand well across the perch on short legs, with thighs and hocks set well back.

Head study of a Princess Parrot in good condition.

Another South Pacific parrot, this Superb Parrot cock shows wonderful color and a well-conformed head.

Zebra Finch cocks. In the Silver shown to the left, the cheek patch contrasts sharply with the dilute color of the head and neck. The Normal coloration is shown below.

A ZEBRA FINCH STANDARD

Perhaps one of the most popular of the finches is the Zebra, originally from Australia. The Normal Zebra and its many mutations, mainly the Pied, White, Cream, Silver, Chestnut-flanked White, Fawn and Penguin varieties, are most attractive and noticeable at the shows. A judge will consider conformation, condition, deportment, color and presentation of the Zebra.

In *conformation* the Zebra must be a large finch with a well-rounded head. The topline of the head should carry smoothly to the back with no indentation at the neck. The topline to the tip of the tail must be straight. The chest must be well rounded; Zebras that are too chesty lose points. Conformation is worth 40 points.

Condition of the Zebra is next in importance. If there are feathers missing or wing and tail feathers damaged, points will be deducted from the total. Light or thin-feathered Zebras will lose points, as will obese birds. Condition is worth 30 points.

Color is almost as equally important. One should look for vividness of color. Beak color is stressed with as much as ten points. In total color is worth 20 points.

The remaining ten points are given to *deportment* and *presentation*, with the bulk applied to deportment. If the Zebra Finch flutters upon the approach of the judge, then deportment is downgraded.

A SOCIETY FINCH STANDARD

The Society Finch, or Bengalese, was developed in Asia by man. For years the finch has been used as a foster parent for other harder-to-raise finches. Here is a standard for the Society Finch:

Condition—Feather quality is considered; there must be no fringe areas on the wings or the tail; the finch should look neat and clean—20 points.

Above: Cuban Finch showing poor condition along the head and neck.

Facing page: White-breasted Gouldian Finch. Overlong claws are the obvious fault in this bird, which is not in show condition. These factors should outweigh any consideration deriving from the novelty of this variety.

Conformation—Body conformation must show a nicely rounded body and a round head from the tip of the beak to the nape of the neck with as little neck as possible. One major fault of many Societies is a V-shape between the head, and body. The beak must be in proportion to the size of the head, and the eyes must be set well back—20 points.

Tail—The tail should be straight and clean with no loose feathers—10 points.

Wings—They must be compact, set close to the body and just meeting at tips—15 points.

Deportment—Society Finches should be able to hop from perch to perch but when steady should be perched at a 45-degree angle—15 points.

Markings and color—20 points.

Total—100 points

When shown in pairs, accurate matching is a must. A crested bird must be paired with one without a crest. That both birds be of the following colors and markings is appropriate: Chocolate-and-White; Fawn-and-White; Chocolate Self (all chocolate); Fawn Self (all fawn); White Self (all white); Dilute Chocolate-and-White; and Dilute Fawn-and-White.

In the Chocolate-and-White as well as the Fawn-and-White, markings are immaterial. In Chocolate Selfs, the head, wings, tail and halfway down the breast must be dark chocolate, and the lower half of the breast must be light chocolate, with or without fleckings on the center of the back. No white feathers are permissible. In Fawn Selfs, the wing, head, tail and halfway down the breast must be dark fawn, and the lower breast must be light fawn, with or without fleckings on the back. No white feathers are permissible. In Whites, the entire bird must be white; no fawn or chocolate feathers are permissible. In Dilute Chocolate-and-Whites, feather color is pale chestnut and white. In Dilute Fawn-and-Whites, there must be a white ground with diffused fawn color.

Society Finches. Once the appealing pose is put aside, closer inspection indicates that these Chocolate-and-Whites are not marked similarly enough to warrant exhibition as a pair.

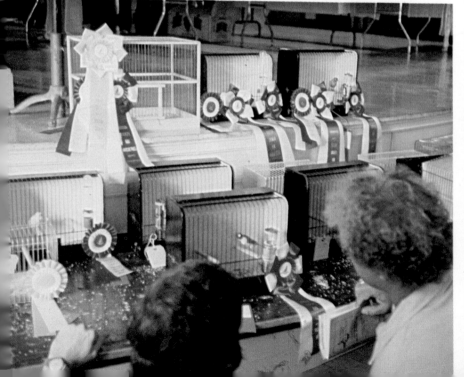

Above: Otherwise very nice, this Senegal Parrot displays a defect called "smuttiness" in the color of the wings.

Facing page, above: Superb Starling, Fourth in Division (Foreign Birds) and First in Section (Softbill and Semi-insectivorous) at the 1981 National Show, exhibited by Evert Gerritsen.
Facing page, below: Some of the finches exhibited at the 1980 B.S.A. Show.

A SOFTBILL STANDARD

A softbill is a bird that prefers to eat insects, fruits and occasional vegetables rather than seeds. Softbill species range from the toucans to the orioles, starlings and mynah birds. One often also sees cardinals exhibited as softbills at the shows. One must consider the usual conformation, condition, color, deportment and presentation when judging the softbills.

In *conformation* one must differentiate the kinds of softbills that are presented. For example, there is a myriad of thrush-like birds. They must conform in head style and body style to the typical thrush. Starlings and mynahs are evaluated otherwise with respect to head structure and posture. A judge must learn these characteristics and become completely familiar with traits of a species or family.

Condition varies little from other categories in that wing and tail feathers must be free from fringing and certainly must not be soiled. Missing feathers and distinct baldness can be a problem and will lose points.

Deportment is more important with the softbills than it is in the finches. Softbills must manage to perch well with no fluttering or severe excitement. The only real exception would be that of the hummingbirds and the like.

Color is assessed for both vividness and pattern. Brightness of color is of obvious value in judging softbills.

Presentation is important to the point that the tail and wing feathers must not be damaged in a cage not suitable for the bird being judged. In any event, one must pay a bit more attention to presentation with softbills than with other categories.

Some softbill judges utilize the judges' scorecard to a great extent. They consider conformation, condition, color and deportment as well as *rarity of species*.

CONCLUSION

Standards are important, particularly once certain kinds of birds become so popular that numerous specimens are exhibited. A standard is developed by an active group of breeders exhibitors and judges of that particular breed of bird. Their interest becomes so keen that they develop a standard in writing. Often the group will secure an artist to portray the ideal specimen. This standard is debated, written and rewritten. Once tried at various shows, active participants may make further changes.

As standards become a fixture, entries increase, and soon that breed which may have only been given a class a few years before now has as a division to itself. Cockatiels, lovebirds, Gloster Canaries, Border Canaries and Budgerigars have all reached this status; Zebra Finches and Bengalese Finches may soon attain it.

Exhibiting can be great fun; it can be extremely rewarding. One can breed birds throughout the year according to the standard, striving toward the ideal. By breeding and exhibiting to the standard, one can achieve great results.

Two closely related conures. The Golden-
capped (*above*) exhibits typical mature coloration, while the coloration of the wing of
the Sun Conure (*facing page*) indicates that
this is still a young bird. Its conformation is
good, however.

INDEX

Numbers set in **bold** refer to illustrations.